Pocket Injection Mold Engineering Standards (2nd EDITION)

This book includes many reference tables and graphics supplying valuable information for injection mold design and engineering.

The book includes mold specification sheets and mold design/engineering for gates, cooling, sprues & runners, runner sizing, ejection, pullbacks & KOs, SPI KO patterns, clamp slots, venting, hydraulic cylinders, slides, alignment, O-rings, SHCSs, support plate & pillars, hot runner considerations, etc.

Also included: mold design checklist, quoting & design direction, tips to best determine shrinkage values for X,Y & Z axis, mold steels and hardness, heat treatment and tempering data, thermal conductivity values, thermal expansion, plating, best surface treatments, surface finish tables, edm roughness table, updated list of common suppliers, and more.

This new 2nd EDITION also includes selected additional reference pages from other APEBOOKS which are related to mold engineering.

See TOC for listing of all subjects covered.

Copyright & ISBN

Contributions include miscellaneous figures/graphics used with permission from Husky Injection Molding Systems Ltd., Canada. Graphics also used with permission from PCS Company, Fraser, MI, USA.

Library of Congress, copyright number: TX0005080970

ISBN-13: 978-1466450844 .... ISBN-10: 1466450843

About the Author

The author - Jay Carender:
- Hands on processing skills,
- Mold build/mold design & engineering knowledge,
- Degree in Mechanical Engineering from GMI (General Motors Institute of Technology ... now known as Kettering University),
- DOE Training from Stat-Ease, Inc.,
- SPC Training from University of Tennessee Management Development Center,
- SPC Training from ASQC,
- Black Belt from AIT (Advanced Integrated Technologies),
- Productivity and Quality Improvement Training by Dr. Deming,
- Mold Design & Advanced Mold Design Training from New York University ... taught by John Klees Enterprises,
- Multiple courses on processing from RJG Industries, Inc.,
- Owner of Advanced Process Engineering - company started in 1990 to create and market pocket sized reference booklets for injection molding industry. Six booklets written & published along with other training manuals.

The hands on experience is from various fortune 500 companies performing injection molding. The extensive experience includes:
- Hot runner molding,
- High cavitation molds,
- High speed molding with cycle times less than 5 seconds,
- Stack molds, unscrewing molds, core pulls, slides, close tolerance parts,
- Engineering resins,
- SPC,
- Statistics,
- DOE to effect process improvement and dimensional nominalization.

Other APEBOOKS

Injection Molding Reference Guide, 4th Ed.
contains basic part design, trig tables, calculation for thermal expansion w/ coeffs, SHCS data, torque specs, shrink data, cooling equation, mold debug guidelines, melt index data, resin density data, many tables of process guidelines, process development techniques, calculating heat load & water flow requirements, pipe data, conversion factors, transformer & motor current, PM & safety, basic statistics, equip selection guidelines and more.

Injection Molding Troubleshooting Guide, 3rd Ed.
contains troubleshooting tips/solutions for many injection molding defects, intro to DOE, discussion of VPT and Decoupled Molding[SM] techniques (SM - RJG, Inc). This 3rd ED. includes many select pages from other APEBOOKS which are applicable to process set up and troubleshooting such as sources of variation and root cause analysis.

Math Skills for Injection Molding, 2nd Ed.
contains intro to basic algebra, using conversion factors, percentages, ratios, proportions, trig as needed for draft and tapers, trig tables, thermal expansion calculations, calculate shrinkage, determining part cost, understanding efficiency and utilization, intensification ratios and clamp tonnage, projected area, residence time, cooling time, interpolation, heat load, Cp, Cpk, Pp, Ppk, correlation, math equations and samples for calculating piezo and strain gage transducer full scale pressure, and more.

Pocket Injection Mold Engineering Standards, 2nd Ed.
mold spec sheets, quoting & design direction, shrinkage, mold steels and hardness, heat treatment, thermal conductivity, thermal expansion, plating, surface finish tables, cooling design guidelines, gate designs, runner sizing, venting, sprue pullers, sucker pins, ejection, slides, support pillars, alignment guidelines, O-ring guidelines, hot runner info, torque specs, trig tables and more.

Managing Variation for Injection Molding, 2nd Ed.
understand & quantify variation, 6 sigma techniques, Cpk, Ppk, Z-score math, correlation, single & multi regression analysis to create predictive equations, DOEs, ANOVA, components of variance - how to quantify % each, MSE & Gage R&R, SQC, control charts, real time SPC, process mapping, process qualification & validation, FMEA, nominalization, molding techniques to reduce variation, and more.

Basic Statistics and SPC
basic statistics for operators, inspectors, set-ups, etc to prepare personnel for more effective SPC techniques, includes training on understanding SPC control charts (variable control charts - Xbar & R; Xbar & MR and attribute control charts - p, np, c, u charts ... how to compute control limits). Also included: what is variation, cause & effect diagrams, root cause analysis, histograms, pareto analysis and Gage R&R. Discusses inherent problems w/ X-bar & R charts as used in injection molding; explains better sub-grouping strategy.

The Advanced Process Engineering Guide
a compilation of the first five APEBOOKS in one book ... coming soon!

For questions or comments,
contact Jay Carender and Advanced Process Engineering at: advproeng@gmail.com

Table of Contents

Table of Contents (continued)

Mold Specification Sheet and RFQ

The RFQ should include considerable data that helps achieve success with two main objectives:

- Part Design & Function Requirements
- Manufacturing & Productivity Requirements

The Excel® form on opposite page can be constructed and used to communicate mold specifications and requests for quote (RFQ). The check boxes might be indexed to a different Excel sheet to help automate the quotation process – simple Excel functions.

A better apples to apples quoting process can be realized if the customer actually can create such a RFQ so that everyone is quoting the same thing. Completing such a sheet may be a collaboration between the project engineer and mold builder.

Some mold builders may not be able to perform all the services listed: such as the DOE after mold sample and debug.

In today's CAD rich environment ... some mold builders may balk at supplying detailed mold designs, but these can be a good reference later. The resistance stems from many mold builders working direct from CAD to CAM and may not need detailed drawings for themselves; thus, the mold design time is reduced if detailed drawings are not generated (CAD = computer aided design & CAM = computer aided manufacturing ... a 3D part model may be processed to then drive cutter paths, etc).

The project engineer should be part of this RFQ and supply the following as attachments to satisfy the aforementioned Part Design & Function Requirements:

1. Detailed part drawing (and 3D database) listing all pertinent CTFs, design features, quality requirements, surface finish, resin requirements, etc.
2. Gate vestige requirements: acceptable height, below flush, ... cosmetic allowances, etc (if not listed on detailed part drawing).
3. Molded part shrink information, ID responsibility for shrink, review how it will be determined, discuss steel safe construction ... if needed

    in many cases, the customer is better able to determine the shrink rate, than is the mold builder[1]

---

[1] Careful thought is needed here ... See also pages on shrinkage in this booklet which discuss this often troublesome determination. Make sure this responsibility is clearly defined.

If not for shrinkage, injection molding would be easy, planning the dimensional shift resulting from shrinkage is what makes injection molding a challenge for parts having critical dimensions. Keep in mind that each part feature and dimension may actually have a different shrink rate due to variables like cooling rate (in part and mold), process pressure gradient thru out molded part, distance from gate, venting, etc. The mold builder typically can accept one shrink rate for each the X,Y & Z directions in part. The engineer must pick the shrink rate(s) that satisfy the part tolerances or adjust the tolerances as needed.

Mold Specification Sheet ... RFQ Example

<table>
<tr><td colspan="4" align="center">Injection Mold
Request for Quote & Specification Sheet</td></tr>
<tr><td>Date:</td><td>6/4/1999</td><td>Quote Number:</td><td>4455</td></tr>
<tr><td>Customer:</td><td>Advanced Proc Eng'rg</td><td>Part Name:</td><td>Tube</td></tr>
<tr><td>Customer Contact:</td><td>Jay Carender</td><td>Part Number/Rev:</td><td>234-448</td></tr>
<tr><td>Customer Location:</td><td>Phoenix, AZ</td><td>Resin:</td><td>PET EN012</td></tr>
<tr><td>Customer Phone Number:</td><td>480-585-4192</td><td>Mold Cavitation:</td><td>16</td></tr>
<tr><td>Customer Fax Number:</td><td>480-585-4193</td><td>Number of ALL Dims:</td><td>32</td></tr>
<tr><td>Customer email:</td><td>advproeng@aol.com</td><td>Number of Critical Dims:</td><td>6</td></tr>
</table>

SERVICES REQUIRED

DESIGN	BUILD[1]	SAMPLE[2]	METROLOGY
☑ Mold Design	☑ Mold Build	☑ Mold Sample & Debug	☑ FAI (1 shot - all dims)
☐ Layout Only	☑ Class 101	☑ Process Sheet	☑ Cpk Analysis & Report
☑ Fully Detailed (2 sets)	☐ Class 102	☑ PC study - 8 hr/16 shot	☑ DOE Analysis & Report
☑ Shrink from Customer	☐ Class 103	☐ PC study - 24 hr/24 shc	☐ Describe in comments below
☐ Shrink from Builder	☐ Prototype	☐ DOE (2 factor)	
NOTE: Shrink from builder may require unit tool design & build; depends on tolerance & resin.	☐ Spares (25%)	☑ DOE (3 factor)	
	☑ Spares (1 stack)	☐ DOE (4 factor)	

1 - Class 101 - 1MM plus cycles (highest quality); class 102 is high quality - less than 1MM; class 103 is medium duty with lesser hardness - cycles less than 500K
2 - Sampling resin to be supplied by customer.

INJECTION MOLDING MACHINE

Manufacturer tons/shot (oz):	Engel 150 T/10.32 oz	Max daylight (inches):	29.13
Platen size H x V (inches):	28.35 x 25.2	Min shut height (inches):	5.91
Tie bar Spacing H x V (inches):	19.69 x 16.54	Ejector Stroke (inches):	5.12

RUNNER TYPE

☐ Full Cold - 2 plate	☐ Cold - 3 plate	☐ Hot to Cold	☑ Full Hot
☐ Full round	☑ Gate Inserts	☐ Hot Sprue only	☑ Husky
☐ Trapezoidal		☑ Manifold Required	☐ MoldMaster
☐ TDB			☐ Other - See Comments Below

TYPE OF GATE

☐ Edge sq or rect	☐ Full round	☐ Banana	☐ Hot Tip
☐ Edge - round	☐ Sub	☑ Hot Valve	☐ Hot Edge

EJECTION

☑ Single stage	☐ Pins	☐ Air poppets reqd	☐ Lifters A-Half
☐ Two stage	☐ Blades	☐ Mechanical slides	☐ Ejector pins A-Half
☐ Spring return	☐ Sleeves	☐ Hydraulic slides	☐ Lifters B-Half
☑ Pullbacks tied to machine	☑ Stripper plate	☐ Cavity splits	☐ Part sticks to/ejects from A half
☑ KO extensions	☐ Ej plate return limit switch	☐ Slide limit switch	☐ Slide mechanical detents

MOLDING SURFACES

	Core	Cavity	Other
Material	S-7	H-13	
Hardness (Rc)	54-56	50-52	
Surface finish	B1	A3	
Texture	draw polish finish TBD	1000 grit draw finish	
Surface treatment			

OPTIONS

☐ Pressure sensors - 1	☐ Runner Shut-offs	☐ Date inserts	☐ Rev level pin or insert
☐ Pressure sensors - 2	☑ Clean hot tips in press	☐ Engraving - Describe below	☐ Insulator boards
☑ Thru hole clamping	☑ Mold filling analysis	☑ Mold Cooling analysis	☐ Pack & warp analysis
☑ Coolant manifolds	☐ Lift rings	☑ HR controller	☑ EOA for robotics

OTHER COMMENTS:

1 - Gate to be plunger style valve gate (straight - no taper) @ 0.048 inches diameter

2 - Use 0.005 in/in as shrink rate

3 - Mold to have hyd eject cyls; need 12 T. force; press only has 6.8; suggest 4 cyls @ 2½ inch diam

Quoting & Design Direction for Manufacturing

Page six discussed RFQ items (1-3) to insure part design & function success, but we also need to satisfy the ability to manufacture this part and/or assembly with acceptable profit and productivity levels.

Certain information has a large impact on price, productivity and maintenance; thus, should also be specified:

4. Cavitation
5. Delivery requirements
6. Number of molds required
7. Drawings required: fully detailed, layouts only, CAD files, etc.
8. Gate type - sub, edge, hot/cold, valve, pin point, etc.
9. Hot, semi-hot (hot to cold) or cold runner (HR system supplier)
10. Mold base steel – #2 or stainless are common choices
11. Mold operation – full auto, semi-auto, hand unload, etc.
12. Surface finish (may be fully specified on product drawing ... or not)
13. Resin process data such as mold & melt temps planned, will regrind be used, will there be contamination, will there be color changes
14. Surface treatments such as platings, coatings, etc.
15. Spare components required and level of interchangeability with or without custom fitting

Other necessary information to provide

16. Molding machine information (detailed)
    platen drawings w/ KO locations, bolt holes, tie bar clearance, etc
    ejector needs (pull backs ... is machine ej plate drilled for pullbacks)
    ejection stroke & pressure limitations
    does machine have specific min/max stroke requirements
    head room considerations for hoist ... lifting mold over tie bars
    core pulls, air circuits for air poppets, valve gate controls, etc
17. Robot EOA information if applicable ... does EOA need to dock w/ mold's parting line ... post mold part handling plans (drop parts, grab w/ robot ... from top or side, etc).
18. Estimated planned cycle – impact on runner size or design ... cooling needs.
19. Cycle expectations
20. Mold life expectations – number of shots (1M, 3M or 5M, etc.; warranty details?)
21. Engraving requirements for mold exterior: mold name/number, number water lines and label as IN vs. OUT, mold weight, air, hydraulics, etc.
22. Mold sampling requirements – test spares? Basic debug or achieve specific cycle and dimensional requirements? DOE, PC studies, etc.
23. Hardware compatibility needs such as water fittings, hydraulic fittings, safety straps, mold half alignment locks, cylinders – air or hydraulic, hoist rings, metric vs english threads, etc.
24. Dimensional compliance required such as steel to use no more than 25% of the part dimension tolerance; responsibility to correct non-compliance of mold dimensions and/or part Cpk requirements?

NOTE: Customer supplied specifications will affect mold cost, and may stifle mold suppliers ability to provide best engineering AND relieve the supplier of responsibility for which you are paying for ... thus, to be considered! The best solution is a collaboration between project engineer and mold builder to determine best mold specification package.

Mold Type/Construction

High quality molds are typically referred to as "class 101" which means the mold will be the highest quality and last a minimum of 1M cycles and typically includes the following:

1. Detailed design to be provided (reviewed and approved by customer prior to mold construction).
2. Mold base to be hardened to minimum of 280 BHN.
3. Molding surfaces to be hardened to a minimum of 48 Rc (inserts stamped with construction material and hardness).
4. Ejection system to be guided.
5. Slides, heel blocks, gibs, etc to be hardened.
6. Mold surfaces which directly form the molded part should have direct water cooling when possible (as opposed to nearby plate cooling which must then cool the insert which cools the molded part – inserts lines are a major impediment with heat transfer).
7. Plates are often stainless steel (SS; preferred) or protected with plating if not stainless to reduce corrosion. Inserts may also receive plating or utilize SS to reduce negative effects of corrosion (Note: The optimum material for certain inserts and their performance requirements may not permit plating). O-ring surfaces should be SS or plated w/ electroless nickel.
8. Mold to include parting line alignment locks (in addition to leader pins).

Other Basic Construction Requirements

- Ejector system should include provisions for ejector plate return before mold closing - springs or preferably threaded holes for pullbacks. Pushback or return pins should also be present.
- All plates should have threaded eyebolt holes on each of the four ends. Components which weigh more than 25 lbs should also have eyebolt holes for handling purposes.
- Support pillars should be sufficient to prevent plate yielding or warping from injection pressure.
- Water lines and fittings should not be located in clamping slots. Consider recessed water and air line fittings.
- All cavities should include engraving to identify molded part cavity ID. Adopt a standard numbering sequence such as counting from top to bottom starting on top back side; each column coming toward operator side would continue the count from top to bottom.
- All gate sizes should be the same. Do not balance mold fill by changing gate size. If mold fill requires artificial balancing; change runner branch diameters. Geometrically balanced equal flow length runner systems are preferred.
- Leader pins should engage its bushing at least ¼" before core enters the cavity or angle pins engage the slide.
- If surface platings or coatings are to be used; compare the application temperature to the tempering temperature for the mold material (core/cavity/etc); see also page 35 in this booklet.
- SHCS below flush to prevent platen damage during mold close under tonnage in molding press.
- Radius or chamfer all exterior plate corners for safety; machine in prybar slots at sides (preferred) or corners.
- Specify that no welding shall take place without notification to and approval by the customer.

Mold Design/Drawings

Drawings

- The mold supplier should supply a plan layout for review and approval prior to starting the detailed design.
- The final mold design should include details with sufficient dimensions to permit making any replacement components without checking the mold or components in advance.
- Detail identification on assembly view should be as follows:

Detail # --- (4 / 13) --- Found on sheet #

- Sheets with details should include a "Detail Information Block" which includes:
  Detail number
  Detail name
  Material
  Hardness
  Quantity required
  Finish
  Surface treatments, coatings or plating required.
- The design or drawing package should include a BOM (bill of materials). The bill of materials should list each component needed to build the mold sufficiently to permit ordering of replacements. Purchased components which must be altered should be identified as "altered" on the BOM.
- When a solid model is supplied: reference information should be listed on mold design as to what filename and rev level was used for the mold design.
- List on drawing or title block the shrink rate used for design.
- Title Block Information:
  Customer Name
  Address & Phone/Fax Numbers of Mold Builder
  Mold Description
  Part Name; Part/Drawing Number; Rev Level
  Resin and Shrink
  Designed By & Date .... Checked By & Date
  Sheet # ....Total Number of Sheets
- Drawings should be full scale whenever possible.
- The mold tolerances should not typically exceed 25% of the part dimension tolerance....this should be specified on RFQ (request for quote) and/or the customer's purchase order.
- Mold dimensions forming critical part dimensions should have the dimension and tolerance enclosed in a rectangle (place note on sheet to identify this).
- Each sheet should have an Engineering Change block listing engineering changes made.
- Sheets 2 & 3 typically show widthwise and lengthwise cross section views showing support pillars; plate thickness; leadering; return pins; ejector system; melt delivery system (cold or hot runner); main parting line, mold height, width & length; etc.
- Sheets 4 & 5 typically show plan views which show tie bar locations and platen bolt hole locations; cavity and core locations; offset corner noted; leader pins; KOs; ejector & return pins; etc.
- Additional sheets may show mold opening details, added explanation of water schematics; misc section views and misc details as mentioned above.

Mold Shrinkage

Plastics are heated during injection molding; plastics expand when heated; the total % expansion contracts during cooling and becomes the average shrinkage.

The design and build requires great skill and precision equipment, but predicting shrinkage is also a huge challenge to the mold builder. If it were not for shrinkage, there would be far more companies interested in building molds, but molds have to make parts the right size; thus requiring effective mold engineering. A mold with a dozen critical dimensions can quite possibly (probably) have a dozen different shrink rate values - a different shrink rate for each dimension. This is because all the following factors do affect shrink rate:

Shrinkage is altered by:
- Inherent plastic material properties: crystalline vs. amorphous, specific heat; thermal diffusivity; thermal conductivity, etc.
- Cooling rate (affected by water line location, water temperature, steel selection, area of steel versus plastic volume to be cooled, etc). Cooling rate affects how long packing can take place and degree of crystallinity with crystalline resins....more crystalline results in a more space efficient, orderly structure which means more volumetric change from the random structure of the amorphous melt (all plastics are amorphous at processing melt temperatures with the exception of LCP – liquid crystal polymer).
- Wall thickness (mainly because it affects cooling rate)
- Flow direction affecting orientation and retained orientation – molecules align during flow (caused by shear and molecular shape). As the molecules cool, some orientation may be retained.
- Process packing pressures affect how many molecules are packed into a given space – the cavity.
- Fill rate affects pressure drop in mold and resulting cavity pressure plus affecting amount of molecular orientation.

As previously mentioned, both cooling rate and flow direction affect shrinkage, it becomes easily understood how there can be many shrink rates in the same part. Parts that are complex in shape result in molds that are complex; thus, the cooling likely is not the same for all areas of mold and part: some areas have closer coolant channels than others. Good mold design minimizes such differences, but there will be different cooling rates thru out the mold.

Some complex parts require cores and cavities in both mold halves; thus, the part may try to stick in both mold halves or the incorrect mold half. Proper mold design can often resolve these problems thru the use of optimum draft angles and surface finish, but sometimes the process must be adjusted in attempt to affect shrinkage. Typical process adjustments used to control which mold half a part sticks to include: pressure adjustments and individual mold half temperature adjustments.

Note also: Shrinkage differentials may result in warpage; shrinkage is more predictable with amorphous resins and much less predictable with crystalline resins.

Calculating Shrink & Cavity Sizing

Shrink rates are listed as inch/inch or as a percent...one is same as other only two decimal places different (percent is "per hundred"; thus, the decimal point is two places to the right: 0.018 in/in equals 1.8 %). A polypropylene ruler that is 12 inches long with the aforementioned shrink rate would require a cavity length that is 12.220 inches long. This is derived from the formula below. Many people in the industry calculate this as follows: 12 X 1.018 = 12.216 inches...this mold cavity is 0.004 inches too short, but who will notice. This math error continues today by some mold builders, but largely goes unnoticed because there is often a part tolerance that permits it and if the tolerance does not permit it, then the error is dismissed as actual shrink being different than predicted. Whether or not it is an error depends on method used to calculate the shrink.

Formulas for Calculating Shrink and Cavity Size

$$\text{Shrinkage} = \frac{(\text{cavity dim - part length})}{\text{cavity dimension}}$$

$$\text{Cavity Dim} = \frac{\text{finished part length}}{(1 - \text{shrink rate})}$$

Method #1 (correct method)

$$\text{cavity} = \frac{\text{finished part length}}{(1 - \text{shrink rate})} = \frac{12.000}{(1 - 0.018)}$$

$$\text{cavity} = \frac{12.000}{0.982}$$

$$\text{cavity} = 12.220 \text{ inches}$$

$$\text{shrink} = \frac{(\text{cavity dim - part length})}{\text{cavity dim}}$$

$$\text{shrink} = \frac{12.220 - 12.000}{12.220}$$

$$\text{shrink} = 0.018 \text{ in/in}$$

Note: The people who are said to calculate shrinkage incorrectly; would typically calculate shrink in the following manner which results in the CORRECT CAVITY SIZE IF AND WHEN SHRINK RATE IS CALCULATED AS FOLLOWS....the problem arises when a resin supplier's shrink rate is used and the supplier used different math to calculate the shrink rate. Typically on ultra critical dimensions, the mold will be left steel safe and tweaked or prototyped in a unit tool...in either case correct answers can be obtained with either method so long as the right methods are used together.

Method #2 (alternate method that can work; see note above)

$$\text{shrink} = \frac{\text{cavity dim}}{\text{part dim}} = \frac{12.220}{12} = 1.01833$$

$$\text{cavity} = 12 \times 1.01833 = 12.220 \text{ inches}$$

Improving Nominalization

The nominalization can be improved by the following:

1. Optimal engineering in terms of predicting the actual plastic shrink rate[1].
2. Perform process corrections using the regression techniques learned in the DOE and regression pages found in the *Managing Variation for Injection Molding, 2nd ED* book (these are very powerful techniques for process correcting dimensional problems).
3. Number one above is better achieved by adopting the following: measurements in different parts of mold should include a <u>balanced sampling of wall thicknesses and distance from gate</u>; this improves chances of including different cavity pressures and different cooling rates (cavity pressure will be reduced as you get farther from the gate & cooling rate depends on part geometry and mold design). Determine an average shrink rate for the X, Y & Z axis directions so that flow orientation is accounted for in the analysis (critical w/ high shrink resins like PE, PP, etc where the flow direction may shrink 30-40% more than crossflow direction). $X_1 - Z_3$ below are shrink rates for location.

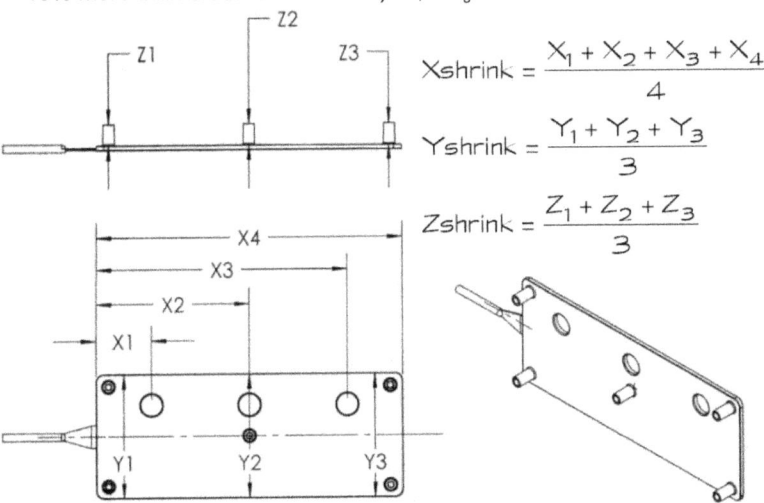

$$Xshrink = \frac{X_1 + X_2 + X_3 + X_4}{4}$$

$$Yshrink = \frac{Y_1 + Y_2 + Y_3}{3}$$

$$Zshrink = \frac{Z_1 + Z_2 + Z_3}{3}$$

5. Verify baseline process is good and typical for new mold (assumes that prototyping is being done via proto or unit tool).
6. Have same person measure both plastic and steel.
7. Have both metrology dept and toolroom do measurements.
8. <u>Reconcile the differences</u> – note that if only one person measures the part or steel they are not wrong no matter what they get ... until the new mold arrives and the actual shrinkage is confirmed for each dimension; thus, take the time to perform parallel shrink determinations and reconcile the differences.
9. If a new mold is built w/o a proto tooling phase:
   a. design/build mold with CTFs as steelsafe w/ planned recut
   b. review similar mold in same resin, but still make CTFs steelsafe
   c. plan sufficient budget to recut or remake some components
10. The project engineer should be involved

---

[1] Remember that every dimension may have a slightly different shrink rate; thus, the best average shrink rate must be determined ... deviations from average must be compared to the part tolerance.

Blank Page

Mold Materials (steel, etc.)

The cost of the mold steel is typically only 5 - 10% of that mold's cost; thus, not the place to be cutting corners. The cost of labor for design and build is the bulk of a mold's cost. Steel selection requires careful consideration of all performance requirements for the given mold.

The upcharge for using stainless plates is not excessive, but the benefit toward eliminating corrosion inside water channels and at O-ring face seals is great if the mold is expected to run a long time or many cycles. Conversely if you have a product that only runs 250,000 pieces and at end of year the mold will be retired, then 4130/4140 plates (aka #2 steel) will be sufficient. If the molding resin is corrosive (e.g. PVC) then stainless steel may be needed for molding surfaces. If the molding resin has certain fillers then abrasive wear resistance is more of a concern. Sometimes the mechanical action of the mold may require certain steel selections so as to permit steel on steel sliding without galling. Molding surfaces of precision optics will need a steel which can be polished to a mirror finish. If the inserts will receive coatings to further enhance performance, then steel characteristics to receive coating or endure coating process must be considered (e.g. coating application temperature versus tempering temperature). Hot runner components often use hot work steel because of their superior properties at elevated temperatures. Very large molds and/or short run molds may use pre-hardened steel (270-350 Brinell) to eliminate the need for additional heat treatment.

When mold steels of high thru hardness are used they are supplied in the soft annealed condition (hardened mold inserts for cores, cavities, misc molding surfaces and gibs, wedge locks, etc are typically hardened to a range of 48 - 62 Rc). They are then rough machined, stress relieved, finish machined and go to heat treatment for hardening and tempering to desired hardness. After this heat treatment, the core or cavity typically must then be finish ground and/or polished. In some applications, there will be additional coatings or textures to further treat the molding surfaces. While steels having higher hardness typically have increased wear resistance, they should be selected when appropriate as those same steels may be more difficult to machine. A D2 or A2 insert will have higher hardness and excellent wear resistance, but will be much more difficult to grind to final shape and size compared to a H13 or S7 insert. Inserts such as Elmax™ and CPM 10V® (AISI A11) and Ferro-Tic® titanium carbide alloy are even more difficult to grind, but afford outstanding wear resistance.

Copper alloys such as Beryllium Copper (BeCu) and Bronze are sometimes used for the high copper content which yields outstanding thermal conductivity (see following pages).

Titanium alloys such as grade #5 are often used due to a couple reasons: low modulus of elasticity and poor thermal conductivity. The low modulus permits material to used as support for HR systems in what will become an interference fit after thermal expansion, and the low modulus permits the material to act something like a spring (to a limit...permanent deformation can take place). The low thermal conductivity also makes it desirable for HR applications such as the aforementioned HR support or "squash" pads, but also in HR transfer seals where the HR tip needs to be isolated from the cooler cavity.

Common Mold Materials

Misc Steel

HRS[1] - Hot Rolled Steel which is typically a low carbon steel (1010/1020) which can be in sheet, rod, plates, etc. This steel might be used for plates on lesser quality molds. 1020 has 0.20% carbon.

CRS[1] - Cold rolled steel is similar to HRS except has slightly better finish and mechanical props at 40-90K psi tensile strength depending on the temper/hardness, but still low carbon steel (1010/1020). Sometimes used for KO rods.

NO. 1[2] - Medium carbon steel (1030) used for some mold plates.

4130/4140 - Is a medium carbon steel typically supplied as pre-hardened (28-34 Rc) and used for plates on medium to higher quality molds (clamp, support, retainer).

NO. 2[2] - Same as 4130/4140 above.

P-20 - A higher grade plate steel that is also used for some cavities and cores in larger molds and/or short run high quality prototype molds. P-20 is a high grade 4140 made to higher purity standards which result in more homogenous structure lending to excellent polish. In addition to plates and large cores/cavs, might be used for HR manifolds. Supplied pre-hardened at 29-36 Rc.

NO. 3[2] - Same as P-20 above.

H13 - Good tool steel having excellent mix of properties when trying to achieve toughness and wear resistance. Tempered at high temperature; thus, can be exposed to high heats in coating process and/or end use application which makes it good for HR components or components mating to same. Typical max hardness is 50-52 Rc. Large cores and cavities, whereby post machining heat treatment is not practical, can be machined from pre-hardened H13 at 44 Rc.

NO. 5[2] - Same as H13 above (fully annealed at 13 - 20 Rc).

414 SS[3] - Stainless steel pre-hardened to 30 Rc; used for cores and cavities to avoid heat treatment and final grind or polish. Has slightly better corrosion resistance than 420 SS.

420 SS[3] - Stainless steel which is supplied both as pre-hardened (33-37 Rc) for use in plates and holder blocks and as fully annealed (14-23 Rc) for use as stock to machine precision cores and cavities from and subsequent hardening up to 50-52 Rc (depending on supplier...some 420 SS should not exceed 50 Rc...check your supplier recommendations). Highest quality class 101 molds use 420 SS plate with various tool steel cores and cavity inserts.

440 SS - Stainless steel with better wear resistance and compressive strength than 420 SS but slightly lesser toughness.

NO. 6[2] - Same as fully annealed 420 SS above.

NO. 7[2] - Same prehardened 420 SS above.

O1 - Used for gibs, slides and wear plates can be hardened as 58-62 Rc. Very low toughness, but very high compressive strength. Not used for cores or cavities.

O6 - Same as O1 above but has free graphite in microstructure providing excellent lubricity. Hardened as 58-60 Rc. Slightly less compressive strength. Not used for cores or cavities.

S7 - Similar to H13, but can be hardened to 54-56 Rc. Slightly better wear resistance and compressive strength than H13, but slightly less toughness. Used for cores, cavities and stripper rings.

A2 - Used for small cores or cavities (inserts) having outstanding wear resistance (better than A6). Hardened at 56-58 Rc. With great wear resistance comes good compressive strength, but poor toughness.

A6 - Similar to A2, but slightly better toughness and slightly less wear resistance. Still excellent compressive strength and wear resistance as needed for small cores and cavities or inserts. Hardness used at 56-58 Rc.

A10 - Used for ejector blades, gibs, interlocks and wedges. This steel also has free graphite which results in outstanding wear resistance. Hardness used at 58-60 Rc.

D2 - Outstanding wear resistance, but has low toughness (brittle). Used for cores, cavities and gate or runner inserts needing superior wear resistance. Avoid sharp corners due to brittle nature. Hardness used at 56-58 Rc.

M2 - A high speed steel used for small core pins, sucker pins, ejector pins and ejector blades. Very low toughness, but very high wear resistance and compressive strength. Hardness used at 60-62 Rc.

Powder Metallurgy Steels (PM)

CPM 10V® - This is a steel made from PM process. This material has better wear resistance than D2 (so much so that is hard to finish grind in hardened state). Excellent for gate inserts, etc where outstanding wear resistance is needed. Hardness used at 60-62 Rc. Supplied by Crucible - address found on page 102.

Elmax™ [4] - Also made from PM process. Excellent wear resistance; good for small parts and inserts.

ASP 23 - Also made from PM process. Excellent wear resistance; good for small parts and inserts. Hardness used at 60-64 Rc.

Non-Ferrous Materials

6160 Al - Marine grade of aluminum. K[6] value of 99 Btu/hr ft °F.

7075 Al - Aircraft grade of aluminum. K value of 70 Btu/hr ft °F.

Protherm® [5] - High thermal conductivity BeCu (Beryllium Copper); hardness at 96 Rb (@17 Rc) and thermal conductivity at 146 Btu/hr ft °F).

MoldMax® [5] - High strength BeCu (2% Be) having a hardness up to 40 Rc, but lesser thermal conductivity at 60 Btu/hr ft °F (when hardness is 30 Rc the thermal conductivity is 75 Btu/hr ft °F).

Ampco Bronze Alloys - Often used for bushing and bearing surfaces, wear plates, stripper sleeves and high thermal conductive inserts.

Ampco 18 - Aluminum bronze @ 92 Rb & K[6] of 36 Btu/hr ft °F.

Ampco 21W - Aluminum bronze @ 29 Rc & K of 25 Btu/hr ft °F.

Ampco 22W - Aluminum bronze @ 35 Rc & K of 25 Btu/hr ft °F.

Ampco 940 - Nickel-silicon-chromium copper @ 94 Rb & K value of 125 Btu/hr ft °F.

Other

Lamina™ - Wear plates made from C1018 CRS electroplated with a bronze alloy that is @ 89% copper and 11% tin with a hardness of 24-28 Rc. The bronze thickness is typically 0.008 to 0.010 inches. Self lubricating plate also available which have oil impregnated graphite insert plugs.

---

[1] Common acronym used for many steels beyond that described here.
[2] Number designations used by some mold base/plate suppliers such as D-M-E.
[3] Stainless steels mated to sliding surfaces sometimes have galling problems; take care to design proper clearances. 440 SS is best SS for galling resistance.
[4] Supplied by Bohler-Uddeholm (see page 102).
[5] Manufactured by Brush-Wellman (see page 102) & supplied by various sources.
[6] "K" denotes thermal conductivity.

Tool Steel Comparison Chart

TOOL STEEL COMPARISON CHART
(higher numbers are more favorable for property listed)

AISI Desig. Description	Typical Rc	Hardening Temp (°F)	Tempering Temp (°F)	Wear Resistance	Toughness	Compressive Strength	Corrosion Resistance	Thermal Conductivity	Hobbability	Machinability	Polishability	Heat Treatability	Weldability
4140	30-36	1500	1200	2	8	4	1	5	1	6	5	10	4
P20	30-36	1600	1100	2	9	4	2	5	1	6	8	10	4
414SS	30-34	1550	750	3	9	4	7	2	1	4	9	10	4
420SS	35-40	1885	1050	3	9	4	6	2	1	4	9	10	4
P5	59-61	1575	450	8	6	6	2	3	9	10	7	6	9
P6	58-60	1475	425	8	7	6	3	3	8	10	7	6	8
O1	58-62	1475	475	8	3	9	1	5	5	8	8	7	2
O6	58-60	1475	500	8	4	8	1	5	7	10	5	6	2
H13	50-52	1875	1000	6	7	7	3	4	6	9	5	6	2
S7	54-56	1725	550	7	8	8	3	4	6	9	8	8	3
A2	56-58	1750	1000	9	3	9	3	4	4	8	7	9	2
A6	56-58	1600	450	8	4	8	2	5	5	10	7	7	4
A10	58-60	1475		9	5	9	2	5	5	8	6	7	2
D2	56-58	1850	950	10	3	8	4	2	4	4	6	9	1
420SS	50-52	1885	480	6	6	6	7	2	4	7	10	8	6
440SS	56-58	1900	425	8	3	8	8	2	3	6	9	7	4
M2	60-62	2225	1125	10	2	10	3	3	2	4	6	8	2
ASP23	61-63			10	5	10	4	3	1	4	7	8	2
BECU (2%)	36-42	625	NR	1	1	2	6	9	10	10	9	7	7
BECU (2%)	26-30		NR	1	1	2	6	9.5	10	10	9	7	7
BECU (0.5%)	20-24	900	NR	1	1	1	7	10	10	10	9	7	9

Note: Hardening & tempering temps are approx. and not intended for heat treatment, but only to reference when hardness may be altered (i.e. some surface treatments may use elevated temps .... if application temp is higher than tempering temp: then hardness may be compromised).

Thermal Conductivity Graphs

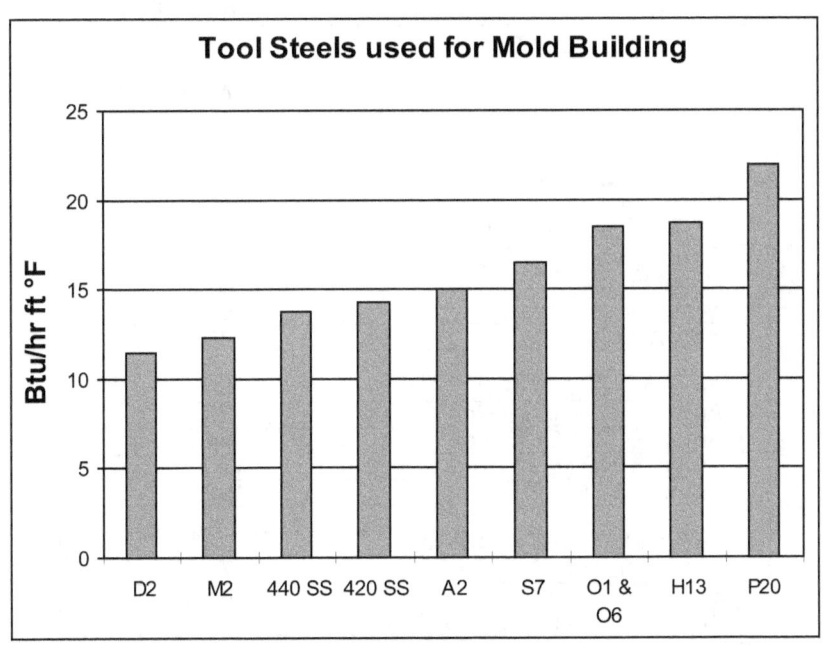

Tool Steels used for Mold Building

Btu/hr ft °F

D2, M2, 440 SS, 420 SS, A2, S7, O1 & O6, H13, P20

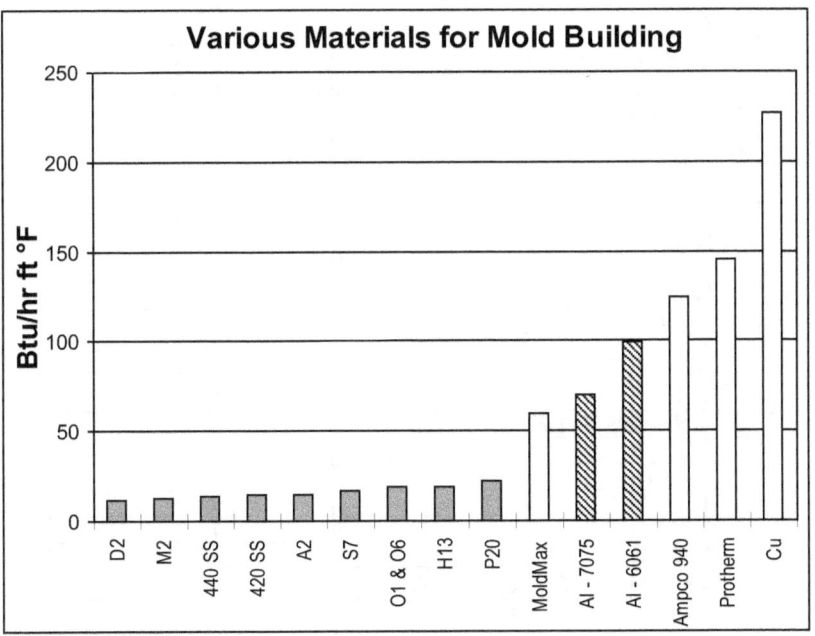

Various Materials for Mold Building

Btu/hr ft °F

D2, M2, 440 SS, 420 SS, A2, S7, O1 & O6, H13, P20, MoldMax, Al - 7075, Al - 6061, Ampco 940, Protherm, Cu

Thermal Conductivity

METAL	Btu/hr ft °F
ALUMINUM ALLOYS:	
1100	137.9
2024	108.9
6061	99.2
7075	70.1
COPPER AND ALLOYS:	
Pure copper	227.6
Free machining (1%Pb)	222.5
Cartridge brass (70%)	70.2
Naval brass	67.7
Manganese bronze	62.9
Phosphor bronze	41.1
Ampco® 18 @ 92 Rb	31.4
Ampco® 21W @ 29 Rc	26.6
Ampco® 22W @ 35 Rc	24.2
Ampco® 940 @ 94 Rb	125.0
Ampco® 97 @ 77 Rb	190.0
BeCu Moldmax® @ 40 Rc	60.5
BeCu Moldmax® @ 30 Rc	75.5
BeCu Protherm® @ 96 Rb	145.6
(BeCu alloys from Brush-Wellman)	
IRON:	
Pure iron	43.0
Cast iron	27.1
Low carbon steel alloys	30.0
High carbon steel alloys	26.1
TOOL STEEL:	
A-2	15.0
A-6	15.0
A-11 (CPM 10V®)	12.4
D-2	11.4
H-13	14.2
O-1	18.5
O-6	18.5
P-20	16.8
S-7	16.5
M-2	12.3
T-15	12.1
STAINLESS STEELS:	
303 SS	9.7
410 SS	14.5
414 SS	14.3
420 SS	14.4
440 SS	14.0
TITANIUM ALLOY (Grade 5)	4.5
TITANIUM CARBIDE	
Ferro-Tic® Grade CM	15.7

Hardness Conversion Table

BRINELL	ROCKWELL B 100 KG. LOAD	ROCKWELL C 150 KG. LOAD	TENSILE STRENGTH (1000 PSI) APPROX.
745		65.3	
712		—	
682		61.7	
653		60	
627		58.7	
601		57.3	
578		56	
555		54.7	298
534		53.5	288
514		52.1	274
495		51.6	269
477		50.3	258
461		48.8	244
444		47.2	231
429		45.7	219
415		44.5	212
401		43.1	202
388		41.8	193
375		40.4	184
363		39.1	177
352	(110)	37.9	171
341	(109)	36.6	164
331	(108.5)	35.5	159
321	(108)	34.3	154
311	(107.5)	33.1	149
302	(107)	32.1	146
293	(106)	30.9	141
285	(105.5)	29.9	138
277	(104.5)	28.8	134
269	(104)	27.6	130
262	103	26.6	127
255	102	25.4	123
248	101	24.2	120
241	100	22.8	116
235	99	21.7	114
229	98.2	20.5	111
223	97.3	(18.8)	108
217	96.4	(17.5)	105
212	95.5	(16.0)	102
207	94.6	(15.2)	100
201	93.8	(13.8)	98
197	92.8	(12.7)	95
192	91.9	(11.5)	93
187	90.7	(10.0)	90
183	90	(9.0)	89
179	89	(8.0)	87
174	87.8	(6.4)	85
170	86.8	(5.4)	83
167	86	(4.4)	81
163	85	(3.3)	79
156	82.9	(0.9)	76
149	80.8		73
143	78.7		71
137	76.4		67
131	74		65
126	72		63
121	69.8		60
116	67.6		58
111	65.7		56

VALUES IN ( ) ARE BEYOND NORMAL RANGE

Calculating Thermal Expansion

This formula is used to calculate thermal expansion: $\delta_t = \alpha \, (\Delta T) \, L$
Where the symbols mean the following:
δ_t is elongation due to temp change,
α is the coefficient of thermal expansion,
ΔT is temperature change ($T_{HIGH} - T_{LOW}$),
L is the length subjected to the expansion.

Example: The leader pins on a mold are 18.75 inches apart. The "A" half is heated from 70° F to 185° F and to 165° F on the ejector half; if moldbase is steel, use 6.6 $\times 10^{-6}$ or 6.6 X 0.000001 as the coefficient of expansion or use table if specific steel is known.

The expansion between leader pin centers (in ejector half) is calculated as follows:
δ_t = 6.6 X 0.000001 X (165 - 70) X 18.75
δ_t = 0.0000066 X 95 X 18.75
δ_t = 0.0117 inches

The bushing holes in "A" half will expand even more at 0.01423 inches since they are in hotter plates (185° instead of 165° F). This yields a 0.0025 inch differential in leader pin to bushing spacing ... this is about the maximum that can be tolerated unless the bushing holes are worn.

Note also that the shut height of this mold is 11.751 inches. This will grow to be 11.759 inches (0.008 inches increase).
δ_r = 6.6 x10⁻⁶ x (185 -70) x 4.563 + 6.6 X10⁻⁶ x (165 -70) X 7.188
δ_r = 0.0000066 x 115 x 4.563 + 0.0000066 X 95 X 7.188
δ_r = 0.003463 + 0.004507 inches = 0.008 inches.

This increase may be enough to throw off the mold protection setup if set very close; thus, requiring the clamp lockup position to be readjusted after temperature equilibrium is established.

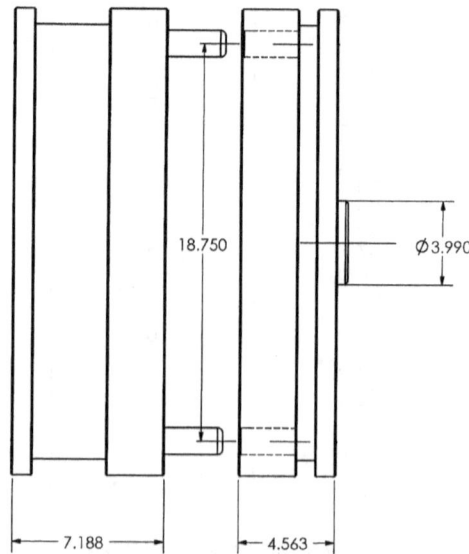

18.750

Ø3.990

7.188

4.563

Thermal Expansion Coefficients - Mold Materials

μ = micro....multiply X .000001; μ in/in from 68 °F to

TYPE	200 °F	400 °F	800 °F
1020	6.5	6.7	7.1
4140	6.8	7.1	
6150	6.8	7.1	7.4
W1	5.8	6.1	7.3
W2	7.4		8.0
S1	6.9	7.0	7.5
S5	6.4		7.0
S6	6.4		7.0
S7	6.8	7.0	7.4
O1	5.8	5.9	7.1
O2	6.2	7.0	7.7
A2	5.8	5.9	7.2
A6	6.4	6.9	7.5
D2	5.6	5.7	6.6
D3	6.3	6.5	7.2
D4	6.2		6.9
H10	6.1		6.8
H11	6.2	6.9	7.1
H13	5.8	6.4	6.8
H14	6.1		
H19	6.1	6.1	6.7
H21	6.9	7.0	7.2
H22	6.1		6.4
T1	5.3	5.4	6.2
T5	6.2		
T15		5.5	6.1
M1		5.9	6.3
M2	5.6	5.2	6.2
M3			6.4
M4			6.4
M7		5.3	6.4
M10			6.1
L2	7.4		8.0
L6	6.3	7.0	7.0
P2	7.0		7.6
P20	6.5		7.1
SS 303,4	9.6	9.9	
SS 316	8.8	9.0	
SS 414	5.8	6.1	
SS 420	5.7	6.0	
SS 440	5.7		
Ti alpha alloy	4.6		4.8
Ti alpha-beta	5.0	5.1	5.2
Ti beta alloy	5.2	5.4	5.6
Ampco 18, 21, 22	9.0		
Ampco 940	9.7		
FREE MACH Cu	9.8		
BeCu (2%)	9.7		
BeCu (0.5%)	9.8		
6061 Aluminum	13.1		
INCONEL	6.4		

Surface Finish Comparison Table

RMS Micrometre (µM)	RMS Microinch (µIN)	Charmilles (LOG #)	SPI FINISH OLD	SPI FINISH CURRENT
.00 - .03	0 - 1		1	A1
.03 - .05	1 - 2		2	A3
.05 - .08	2 - 3			
0.1	4	0		
0.14	5	2		
0.16	6.4	4		
0.18	7.2	5	3	B3
0.2	8	6		
0.25	10	8		
0.28	11.2	9		C3
0.32	12.8	10	4	
0.4	16	12	(280 stone)	
0.45	18	13		
0.5	20	14		
0.56	22.4	15		
0.63	25.2	16		
0.7	28	17	5	D2
0.8	32	18	5	
0.9	36	19		
1	40	20		
1.12	44.8	21		
1.26	50.4	22		
1.4	56	23		
1.6	64	24		
1.8	72	25		
2	80	26		
2.2	88	27		
2.5	100	28		
2.8	112	29		
3.2	128	30		
3.5	140	31		
4	160	32	6	D3
4.5	180	33	6	
5	200	34		
5.6	224	35		
6.3	252	36		
7	280	37		
8	320	38		
9	360	39		
10	400	40		
11.2	448	41		

NOTE: RMS (root mean square) is approx 11% higher than Ra or AA - arithmetic average .... which is average of all areas above and below mean (determined by height and width of peaks and valleys).

Mold Surface Finish Terminology

		Grit	Media
A	1	8000+	# 3 diamond buff
A	2		# 6 diamond buff
A	3	1200+	# 15 diamond buff
B	1	# 600	paper
B	2	# 400	paper
B	3	# 320	paper
C	1	# 600	stone
C	2	# 400	stone
C	3	# 320	stone
D	1	# 11	glass bead (dry blast)
D	2	# 240	oxide (dry blast)
D	3	# 24	oxide (dry blast)

Clamp Slots & Thru Hole Clamping

VARIOUS CLAMP SLOT CONFIGURATIONS

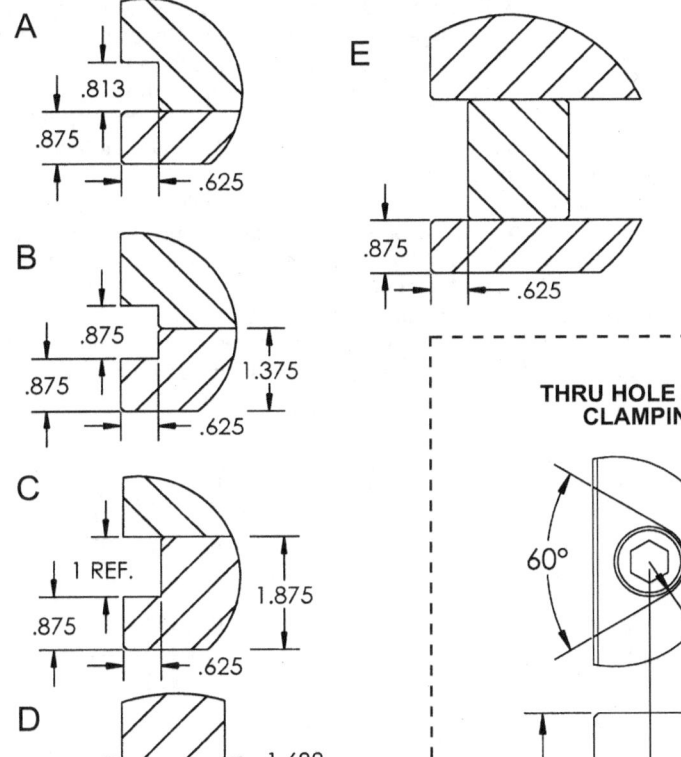

THRU HOLE STYLE CLAMPING[1]

THRU HOLE CLAMPING				
A SHCS Bolt Size	B Thru Hole (in)	C Clearance Radius (min) (in)	D Center to Edge (min) (in)	E Clearance Height[2](min) (in)
1/2-13 x 1.75	0.531	0.40	0.50	2.25
5/8-11 x 2.00	0.656	0.50	0.63	2.75
3/4-10 x 2.00	0.781	0.60	0.75	2.88

[1] Offers convenience and safety advantages;
  mold is bolted direct to platen.
[2] For a 7/8" clamp plate thickness;
  Add ½" for 1.375 plate & use ½" longer SHCS

Cooling: Water Fittings

1. All water fittings and inserts should be made of stainless steel or brass.
2. The water fittings can be selected as follows. To accomplish minimal restriction of coolant flow, avoid use of 200 series fittings except where space available mandates such use. Large molds can use large fittings such as the Cam & Groove fittings or large quick disconnects available from Hansen (Hansen series ST couplings ... find address and web info on p. 102).
3. DME #JP-352 and JPF-0352 or equivalent fittings (Parker #s: PN352 & PN352F) should be used for ¼" NPT water lines. Note counter bore depth and length from table below.
4. When fittings are used on pipe extensions, they should be brazed or silver soldered to the pipe to permit assembly and disassembly as a unit. One piece nipple extensions can be purchased which eliminates need for brazing.
5. Water fittings should be on the operator side or the back side of the mold. Water fittings should not be located on the top of the mold; but when necessary, then mill ½" wide x 1/16" deep grooves across the top and down the sides for water runoff. Grooves should not cross through bolt holes or slide pockets.

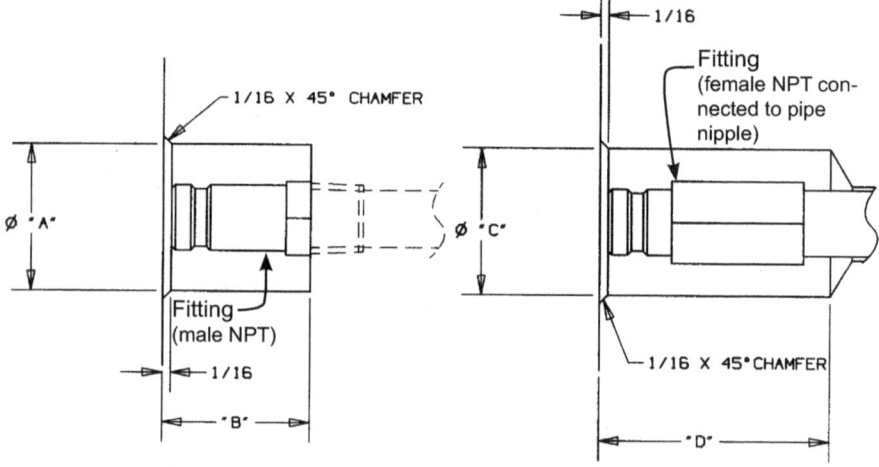

PIPE SIZE	HOLE SIZE[1]	DIAM DIM A	DEPTH DIM B	DIAM DIM C	DEPTH DIM D	ID HOLE[2]	FITTING[3]
1/8" NPT	0.339	0.688	0.688	0.875	1.250	0.250	251
1/4" NPT	0.438	0.844	0.938	0.875	1.500	0.250	252
3/8" NPT	0.578	1.000	0.938	1.000	1.500	0.250	253
1/4" NPT	0.438	1.000	1.090	1.000	1.500	0.375	352
3/8" NPT	0.578	1.000	1.125	1.125	1.500	0.375	353
1/2" NPT	0.718	1.250	1.500	1.375	1.875	0.625	554
3/4" NPT	0.922	1.500	1.563	1.500	1.875	0.625	556
1" NPT	1.156	Cam & Groove Coupling (Brass or SS); Available from McMaster-					
1 1/4" NPT	1.500	Carr. Can also use Hansen ST couplings which are much like					
1 1/2" NPT	1.734	DME or Parker quick disconnects only larger (up to 2½ inch) and					
2" NPT	2.219	higher in cost.					

[1] Based on typical tap drill size which varies per reference chart.
[2] Inside diameter of fitting thru hole.
[3] Automatic shut-off style quick fittings should not be used due to impeded flow by shut-off mechanism.

Cooling Lines Drilled

Algebraic Equation for Thermal Conductivity

$$\text{Time (hrs)} = \frac{L \times H}{K \times A \times \Delta T}$$

L = Distance (ft)   H = Heat (Btu)

A = Area (ft^2) ΔT = Resin melt temp - coolant temp (°F)

K = Thermal conductivity (Btu/hr ft °F)

Two variables which mold designer can control from the five main variables affecting heat transfer above:

1. Distance to coolant from part (L) .... too close and there will be uneven cooling and possibly weak steel condition, but too far away and there will be slower cooling.
2. Core/cav thermal conductivity (K).

Drilled Coolant Lines

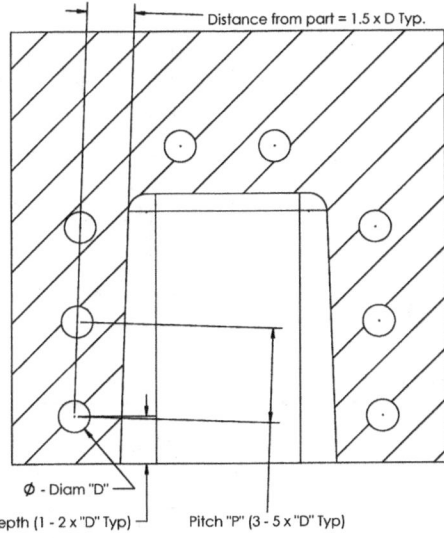

Distance from part = 1.5 x D Typ.

∅ - Diam "D"

Depth (1 - 2 x "D" Typ)   Pitch "P" (3 - 5 x "D" Typ)

Intersecting Coolant Lines

Maintain 1/8 inch minimum clearance; (1/4 inch on holes deeper than 12 inches)

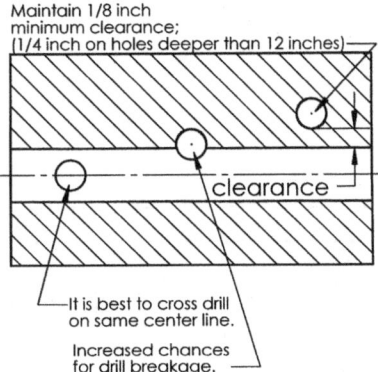

clearance

It is best to cross drill on same center line.

Increased chances for drill breakage.

Baffles

Baffle drops are typically designed whereby several drops are in series; thus, not recommended for small cores of substantial length. The pressure drop in such designs results in low flow rates with large temperature rise in the cooling circuit (bubblers are often the preferred design for many small cores, as each core is in parallel).

Formula for Equivalent Hyd Diameter

$$D_h = \frac{4A}{P}$$

$$A = \frac{\pi d^2}{4}$$

$$P = \pi d$$

A = area of non round hole

P = perimeter of non round hole

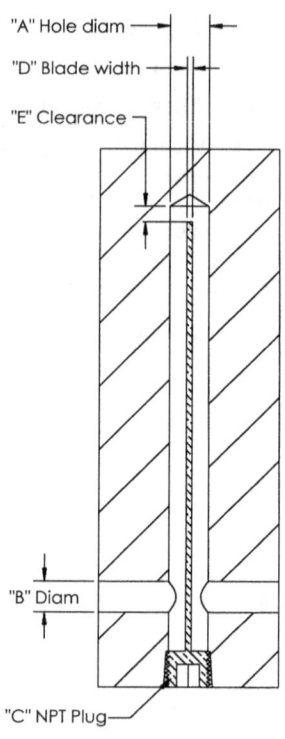

"A" Hole diam

"D" Blade width

"E" Clearance

"B" Diam

"C" NPT Plug

BAFFLE DROP SIZING							
A hole diam inches	**B** (½ round) hyd diam[1] inches	suggested feed line[2] inches	**C** pipe size[3] NPT	Travel per rev inches	Travel per ½ rev inches	**D** blade width inches	**E** gap inches
0.250	0.116	0.250	1/16-27	0.037	0.0185	0.062	0.125
0.312	0.154	0.339	1/8-27	0.037	0.0185	0.062	0.156
0.438	0.212	0.438	1/4-18	0.056	0.028	0.094	0.219
0.562	0.289	0.438	3/8-18	0.056	0.028	0.094	0.281
0.688	0.367	0.438	1/2-14	0.071	0.036	0.094	0.344
0.938	0.502	0.562	3/4-14	0.071	0.036	0.125	0.469
1.125	0.617	0.688	1-11½	0.087	0.043	0.125	0.563

[1] Hydraulic diameter computed using formula above
[2] Needs to be at least equal or greater than hyd diameter of the ½ round.
[3] Hole diameter "A" based on standard blade widths for NPT sizes listed, but optimum drill size for pipe tap may be larger; thus may requiring step drilling

Bubblers

A bubbler consists of a delivery tube located inside of a drilled water passage. Bubblers are useful in cooling long thin cores where drilled passages do not run thru (i.e. blind hole). See figure below for typical layout (in this example, the core is narrow and wide utilizing three bubblers per core, although two would likely work in this application). An annulus is the donut shaped area between two concentric circles (i.e. area of hole ID minus tube OD). There exists at least two methods (listed below) for sizing the optimum relationship between hole ID and tube OD for maximum coolant flow.

Bubbler hole = (tube ID + Tube wall) ÷ 0.707*
*closely approximates equal areas of tube ID and annulus

Bubbler hole = (tube id + tube OD)**
**based on equal hydraulic diameter of tube ID and annulus; hydraulic diameter of tube ID is tube ID, but hydraulic diameter of annulus must be calculated using same formula listed for baffle drops (non-round holes); formula applied to annulus reduces to the above relationship.

NOTES:
1. When bubbler feed tube ID is 0.062" or below, a bag filter (or other suitable filter should be installed on supply side of circuit feeding bubblers).
2. The <u>area</u> of the annulus between tube OD and hole ID will be greater than the area of tube ID. This will create sufficient gap thickness to keep pressure drop to a minimum.
3. Angle cut the tube, just in case construction or installation errors cause it to bottom out with drilled hole; this will help to maintain flow.

A. Bubbler hole inside diameter (sized reflecting allowances for core strength and optimum relationship with tube ID and OD).
B. Coolant channel "out" or "return"; often needs to be larger than "in" since the bubbler tube bisects the hole providing some restriction.
C. Coolant channel "in" or "supply". Area should at least be equal to or greater than sum of tube ID areas (larger to allow for ΔP).
D. Thru hole connecting tube ID to "in".
E. Tube fit length.
F. Tube fit diameter (± 0.001; same as tube OD); line to line friction fit.

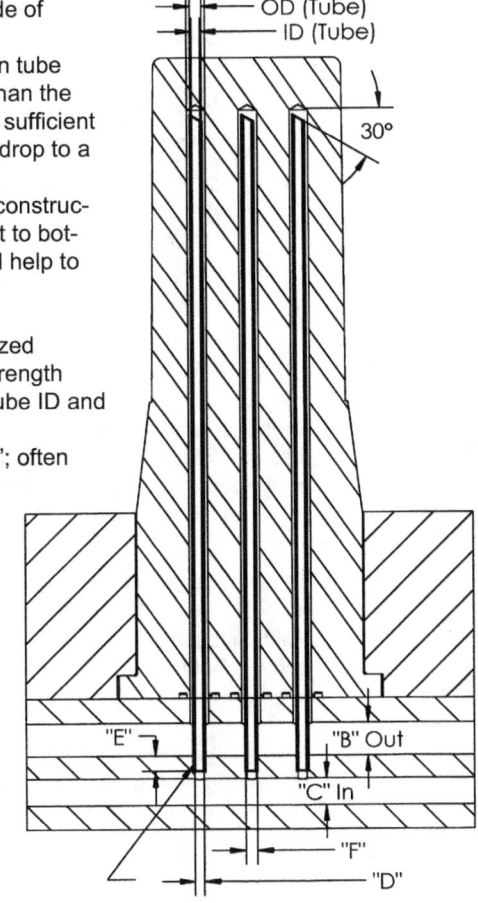

Cooling (an alternative to bubblers & baffles)

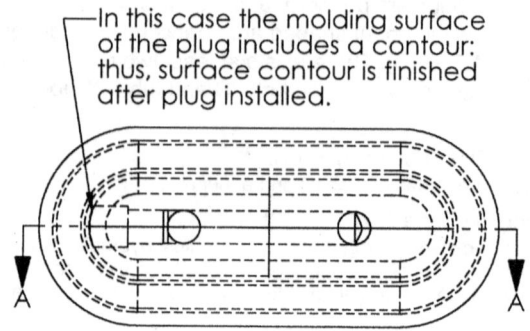

In this case the molding surface of the plug includes a contour: thus, surface contour is finished after plug installed.

A   A

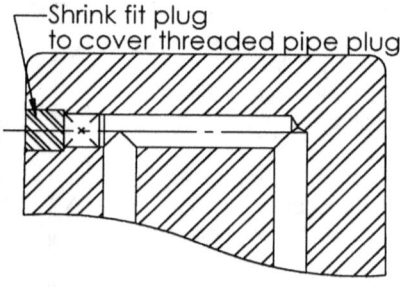

Shrink fit plug to cover threaded pipe plug

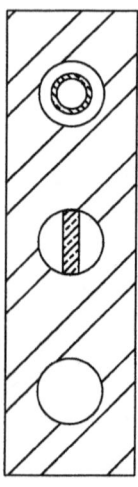

Bubblers must fit the supply and return within same diameter; thus, less flow. Flow will be improved by using thin wall SS or brass tubing (may require custom telescoping construction). This ¼" hole has a 0.156 OD tube w/ 0.116 ID – 0.116 hyd diameter.

Baffles must also fit the supply and return within same diameter. This ¼" hole has a 0.062 wide baffle blade – each half round (less than ½) yields a hyd diameter of 0.116 ....coincidentally the same as bubbler above. Baffles typically are not used in holes below 5/16" diameter.

A full round cooling channel running thru has more than twice the flow versus the aforementioned.....but does require drilling a connecting path which may require a plug which will likely leave a witness line no matter how well constructed.

Cores w/o Cooling & BeCu Cores

Some cores have such a low heat load that they may not require any direct water cooling. The core labeled "A" below, cores out a thru hole in the part. It has good cross sectional area and mass relative to the heat load (even if small diameter with less CSA, would not need direct cooling). It contacts the core which is cooled and is in contact with the cavity which is cooled. This core would be able to conduct its low heat load to surrounding steel and function without direct cooling.

The core labeled "B" has a higher heat load as it picks up heat on face and OD, but the OD is thin wall plastic......this core should be cooled to obtain best molded part appearance (if cosmetics for this blind hole were not important or dimensions not critical, some might elect not to cool this core either, but it would be better to be conservative and design cooling into this core with a baffle or bubbler).

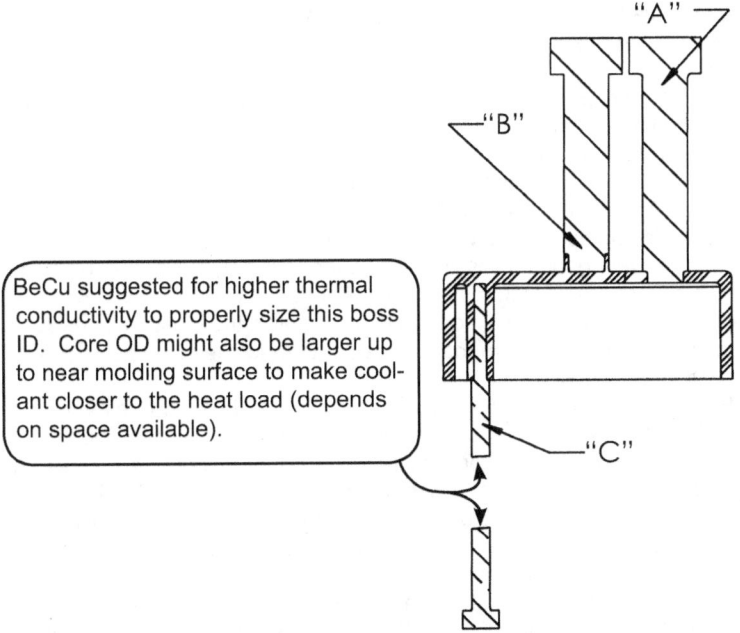

BeCu suggested for higher thermal conductivity to properly size this boss ID. Core OD might also be larger up to near molding surface to make coolant closer to the heat load (depends on space available).

The core labeled "C" is molding the ID for a long boss. The core is only 0.125" OD up to part and 0.090" for the diameter at molding surface. The heat load and distance to cooling is relatively high and the CSA area is low – all heat must pass thru the 0.125" diameter. In this case, it is hard to have direct water cooling in the core; thus, we should consider BeCu or Ampco bronze material to achieve higher thermal conductivity. The 0.125" core diameter could be increased up to the molding surface to permit water closer to the molding surface, but best cycles would require a high conductive core material. Often cores such as this which do not have cooling reach all the way to back clamping plate to permit quick change by unclamping ejector half (mold still clamped onto stationary platen and supported by hoist) then remove a small cover over top core pin......handy when cores become easily damaged.

Surface Coatings & Platings

Mold coatings are typically used to enhance mold performance in one or more of the following areas:

- wear resistance
- corrosion resistance
- improved mold release
- resize components
- combination of the above

The following is partial listing of the more common mold coatings used today for injection molds and components therein.

- CHROMIUM can be applied at a wide range of thicknesses which makes it a good choice for resizing and/or enhancing wear resistance; hardness is 70 Rc and thicknesses from 0.00005 to over 0.030 inches can be achieved. Corrosion resistance is not as good as electroless nickel. Careful location of anodes and electrodes is needed to minimize edge buildup and achieve even thickness.
- ARMOLOY® is a precision hard chrome plating, but has a satin finish or matte finish ...not high gloss shiny as per typical chrome (see also table on next page). Available from Armoloy Corp.
- ELECTROLIZING® is also a precision chrome plating from proprietary solution available only from Electrolizing, Inc.
- NICKEL can also be applied at wide range of thicknesses up to 0.100 inches thick; thus, also very good for resizing. Hardness will only be 40 - 50 Rc. Sometimes referred to as sulfamate nickel. Anodes and electrodes are required and edges can have excessive buildup unlike the electroless nickel. Coating can be applied as more than needed and machined back to size via grind or EDM.
- ELECTROLESS NICKEL (applied w/o electric current; thus, a more even deposition) coating is excellent for corrosion resistance (better corrosion resistance than chrome because of reduced porosity, but less wear resistance). As plated hardness is approx 47– 48 Rc, but can be heat treated to 70 Rc. Thickness is best at 0.0001 to 0.0005 up to 0.002 inches. Good even deposition w/o heavy buildup on corners ... can also coat deep recesses or inside water lines, etc with Electroless Nickel.
- POLY-OND is phosphorus nickel with an overcoating of PTFE (Teflon®). Thickness is typically from 0.0002 – 0.003 inches. Supplier claims benefits include lubricity, wear & corrosion resistance, but lubricity and corrosion resistance are key advantages. Hardness characteristics are same as EN above.
- NICKEL-PTFE is similar to POLY-OND above with main difference being the PTFE is dispersed as small particles into the nickel coating. Author's opinion: This results in a possibly longer retention of the Teflon's lubricity properties, but is a reduced amount of exposed Teflon.....probably slightly less lubricity than POLY-OND, but likely better durability .....differences may be small. (There are many suppliers of Nickel PTFE coatings.) Typical thickness 0.0002 – 0.003 inches. Hardness characteristics are same as EN above. Some claim corrosion resistance improved by the PTFE.

Surface Coatings & Platings (continued)

- TITANIUM NITRIDE (TiN) Physical Vapor Deposition (PVD) of hard coating. This process typically is applied at 930° F; thus, may anneal steel hardness if a reduced application temperature is not requested. Best coating is at normal application temperature; thus, works well with H-13 and steels heat treated and tempered above 930° F. Very hard coating at 82 Rc and 80 millionths thick to 0.0002 inches. Note: there are also other PVD nitride variations, but TiN is most common (gold appearance).
- DiamondBLACK is a ceramic PVD coating, but at a reduced temperature of approx 250° F or less; thus, no adverse effects on steel hardness. Resulting surface hardness is 94 Rc and thickness of 80 millionths having excellent hardness, lubricity and wear resistance.
- DICRONITE® dry lube coatings offer outstanding lubricity (best of coatings listed here) to enhance mold release and/or stop wear, but the longevity or durability of the coating may be less than many other coatings mentioned previously. Dicronite® is a modified tungsten disulfide coating applied at room temperature with a typical thickness of 20 millionths (0.000020).

Coating	Typical Thickness (inches)	Max Thickness (inches)	Hardness (Rc)	Coeff Friction	Approx Cost
Chromium	0.0001 - 0.0005	0.035	70	0.15 - 0.30	$
Armoloy®	0.0002 (chrome)	0.0006	72	0.14 - 0.17	$$$
Electrolizing®	0.0002 (chrome)	0.001	72	0.14 - 0.17	$$$
Nickel	up to 0.100 +	0.100 +	45	0.2 - 0.3	$
Electroless Nickel	0.0001 - 0.002	0.003	48 - 70	0.2 - 0.3	$
Poly-ond®	0.0002 - 0.0005	0.003	48 - 70	0.06 - 0.15	$$
Nickel PTFE	0.0002 - 0.0005	0.003	48 - 70	0.07 - 0.16	$$
TiN	0.00008 - 0.00016	0.00016	82	0.4 - 0.5	$$$$
Diamond BLACK®	0.00008	0.00008	94	0.200	$$$$$
Dicronite®	0.00002	0.00002	NA (30)	0.030	$

NOTES:
1. Most coatings will require EDM surfaces to have the "white layer" removed (layer having surface embrittlement). This done by grinding, polishing or media blasting away. Due to time and cost involved with coatings, it is suggested to consult supplier about optimum surface preparation.
2. Some of the coatings use high temperatures in the coating process and/or post coat process; thus, please compare base metal tempering temperature to coating process temperatures.

Heat Treatment & Hardening

Mold steels are hardened to increase tensile and compressive strengths as well as wear resistance. This prevents abrasive wear and/or damage or distortion from coining or hobbing at the parting line – an overall increased durability. Toughness; however, is typically reduced by heat treatment and it's resulting higher hardness.

Different steels have different resulting combinations of toughness and hardness. In general, steels do not have both great toughness and great compressive strength/ wear resistance at same time. The compressive strength and wear resistance tend to go together, but when compressive strength is high the toughness is usually less and vise versa. H13 steel exhibits a good blend of all properties; thus, it's wide spread use in the mold building industry (see also table on page 18 for comparison of many mold and tool steels/materials).

Heat Treating the steel includes the steel being heated above the critical temperature of 1330° F whereby it changes from pearlite to austenite (carbon goes into solution with the iron). It is held for a time at the elevated temperature, then quenched rapidly. Slow cooling allows the steel to revert back into pearlite; whereas, fast cooling (quench) results in martensite for most steels. These required temperatures are the hardening temperature and vary for different steel types to result in optimum mix of toughness and hardness and yet still avoiding excessive distortion from too much heat. The steel supplier should be consulted to determine optimum heat treat conditions.

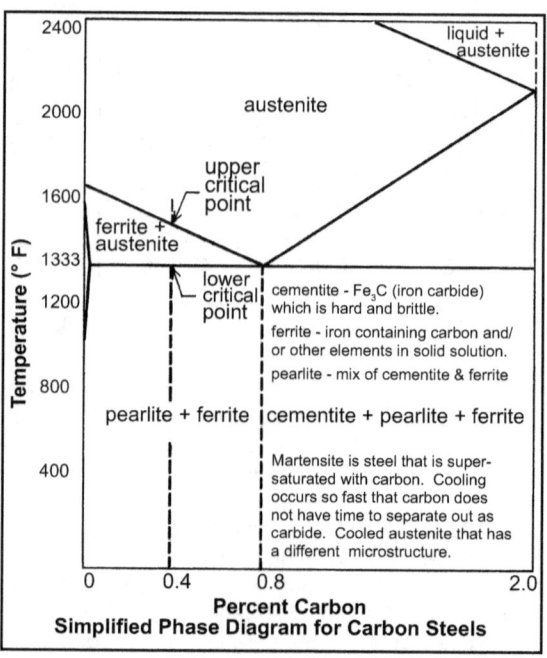

Percent Carbon
Simplified Phase Diagram for Carbon Steels

Heat Treatment Tempering Temperatures

Quenching may take place in air, water (plain or salt), oil, etc. The cooling rate must be faster than the critical cooling rate for the given steel. These critical cooling rates vary for each steel. During quench we quickly bypass the point at which the austenite reverts to pearlite; this results in the formation of hard martensite.

Tempering is done after the initial hardening heat treat and required quench. Tempering is done to both reduce the hardness and increase the toughness; will also relieve many internal stresses and should be done as soon as possible after initial heat treatment. Tempering includes reheating workpiece up to some specified temperature. Higher tempering temperatures result in more martensite being transformed and resulting lower hardnesses; thus, a wide range of hardnesses can be accomplished by varying the tempering temperature. It is important to note that tempering can be done by any form of reheating such as grinding, welding or coating processes after the initial heat treat and temper. Processors should avoid using a torch on mold surfaces to point of heating steel above it's respective tempering temperature for hardness in use. Some high hardness steels may be tempered as low as 350° F; thus, if a mold technician heats that insert up past that 350° F, then the hardness is reduced as well as mold performance. Most mold steels are tempered between 400 and 1100° F. The table below shows tempering temperatures for a sampling of one supplier's mold steels; always obtain data from your specific supplier for best results. The shaded areas are considered most common for injection mold applications, but can be deviated as needed.

TEMPERING TEMPERATURES (relieve stresses, reduce hardness & increase toughness)									
O1		S7		A2		D2		H13	
°F	Rc	°F	Rc	°F	Rc	°F	Rc	°F	Rc
HT	63 - 65	HT	59 - 61	HT	63 - 65	HT	62 - 64	HT	50 - 52
250	63 - 65	300	58 - 60	400	60 - 62	400	60 - 62	1000	50 - 52
300	63 - 65	400	57 - 59	500	59 - 61	500	59 - 61	1050	49 - 51
350	62 - 64	500	55 - 57	600	58 - 60	600	58 - 60	1100	46 - 48
400	61 - 63	600	54 - 56	700	57 - 59	700	57 - 59	1125	40 - 42
450	60 - 62	700	53 - 55	800	57 - 59	800	57 - 59	1150	35 - 37
500	58 - 60	800	52 - 54	900	57 - 59	900	57 - 59		
600	55 - 57	900	51 - 53	1000	56 - 58	1000	54 - 56		
700	51 - 53	1000	50 - 52	1100	50 - 52				
800	48 - 50	1100	49 - 51	1200	42 - 44				
900	43 - 45	1200	37 - 39						
1000	39 - 41	1300	30 - 32						

Annealing is done by reheating workpiece to it's hardening temperature and holding it for given time, then slow cooling to allow the steel to change back into pearlite (slow cooling is critical to fully restore original softness and ductility).

Gates: Straight Tunnel ... aka Sub-Gate

TUNNEL GATES
Offer advantage of automatic degating during mold open.

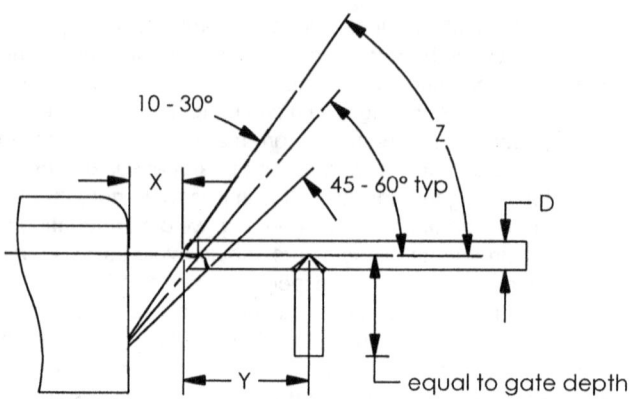

TUNNEL GATE (AKA SUB-GATE)

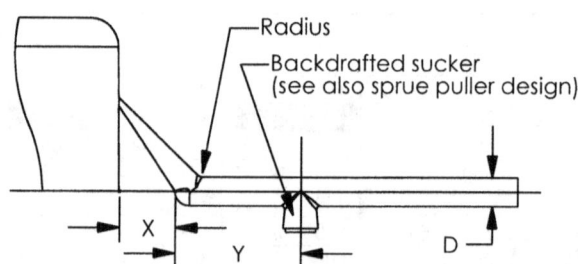

REVERSE TUNNEL GATE W/ SUCKER

PLASTIC TYPE		
	Rigid[1]	Flexible[2]
X	> 0.045	> 0.045
Y[3]	3D or 2X	2d or 1X
Z	≥ 45°	≥ 30°

[1] Rigid resins such as PMMA, PC, PS, PET, ABS, PPO, PA, POM, etc

[2] Flexible resins such as PE, PP, PVC, PUR, TPE, etc

[3] Use max value between two (i.e. max of 3D vs 2X or 2D vs 1X)

Gates: Curved Tunnel ... aka Banana Gate

CURVED TUNNEL GATES

Offer advantage of automatic degating during mold open and ability to gate on bottom side of part. Dimension "L" should be greater than the gate length (Ro x Pi – π ....assumes near 180° arc shape; will be slightly less, but can be a lot less). "D" is typically in range of 0.125" to 0.250". A rectangular profile will work fine and is easy to construct the electrode. It is suggested to make the gate as a split insert to permit in the press cleaning if resin gets stuck in gate (need depends on resin). You will typically curve gate approx 165° then begin added taper to flat where gate orifice will be located. This results in an approx 3° incl taper for t1 to t2 and 6° incl angle for W. Full round designs are also used with included angle at 3-5°. The center for Ro & Ri can be higher to shorten arc length.

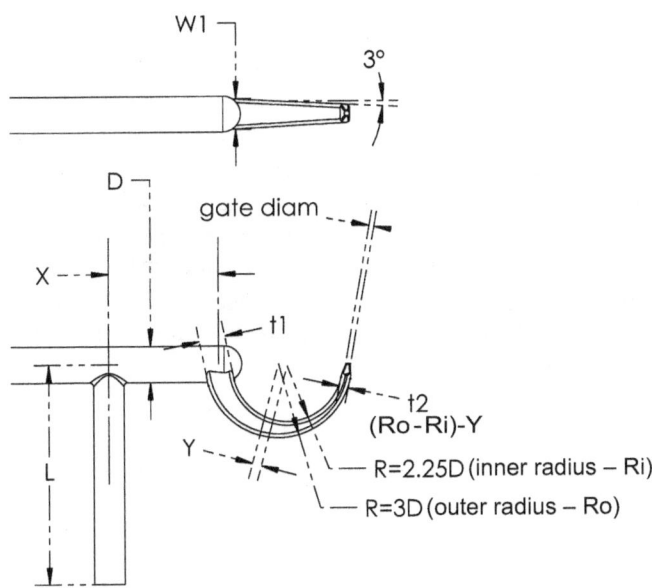

D	W1	t1	Ri	Ro	X	L	Y	t2
spec	= D	0.8 D	2.25 D	3 D	2D	Ro x Pi	0.4 D	calc
0.125	0.125	0.100	0.281	0.375	0.250	1.18	0.050	0.044
0.188	0.188	0.150	0.423	0.564	0.376	1.77	0.075	0.066
0.201	0.201	0.161	0.452	0.603	0.402	1.89	0.080	0.070
0.250	0.250	0.200	0.563	0.750	0.500	2.36	0.100	0.088

Gates: Chisel & Double Chisel

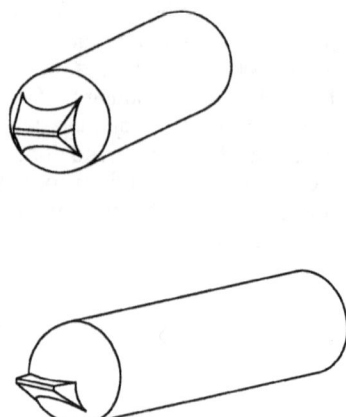

CHISEL & DOUBLE CHISEL GATES
may permit clean break without trimming devices; often used with resins like PS, PPO, PMMA & PC. Can also be used as a sub-gate when small to permit degating during mold open (with appropriate steps in parting line).

As a parting line edge gate, the double chisel design typically is constructed with gate in one mold half plus one 30° approach; then opposite mold half has approach angle only up to a few thousandths of wall.

Gates: Diaphragm & Fan

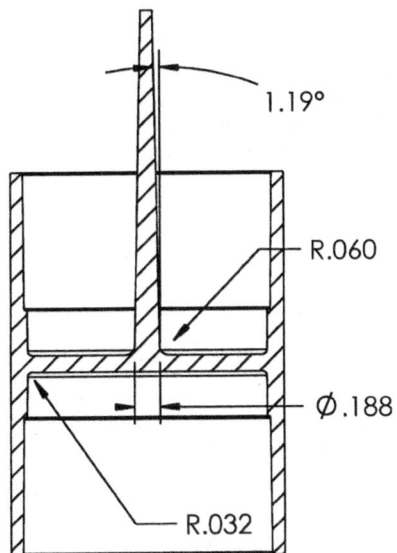

1.19°

R.060

∅.188

R.032

DIAPHRAGM GATES
are used on many round parts such as gears and fittings to preserve roundness, but include disadvantage of the need for post mold degating with punch or drilling operation. The feeder may come from above with three plate construction, direct sprue gated or semi-hot runner design. In this example the feeder taper is same as standard ½" per foot taper used in sprue bushing, but smaller tapers may work and may be required for long drops. In this example the "A" half core molding the fitting will require special cooling such as a BeCu core since only the base may be cooled on OD of core (not shown).

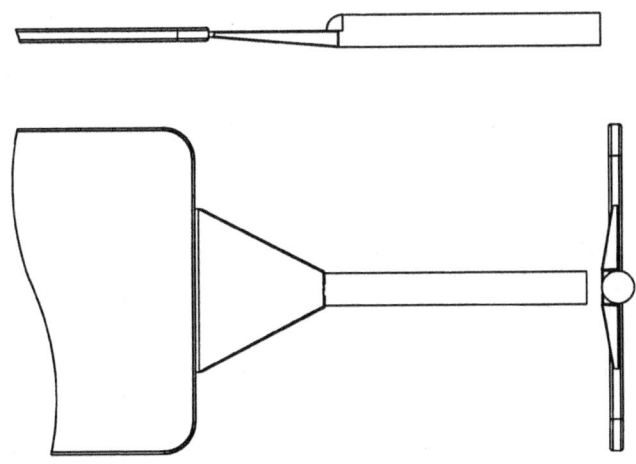

FAN GATES
are used with parts where improved flatness is desired, but also require post mold degating using a shear or milling operation for optimum appearances.

Gates: Edge

EDGE GATES

are very simple gates and easy to machine. These gates can have a large CSA (cross sectional area) to be used on larger parts. Edge gates will require post mold degating. To avoid jetting use a gate depth equal to 50-80% of wall thickness (important when the gate flow does not impinge off core or other mold surface). Suggest start at 50% then can deepen as needed after first sample.

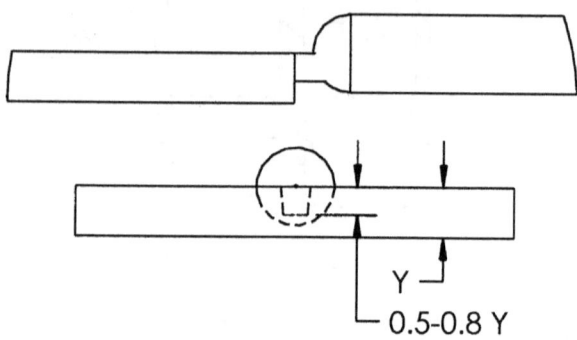

Y

0.5-0.8 Y

Gates: Pinpoint

PIN POINT GATES
typically require three plate mold construction unless a hot tip system is used which
includes both a hot runner system and a hot gate. Cold pin point gates that are three
plate construction need to be designed such that the gate will break at the desired
location – whether that is at the part or a molded post which might be milled off later.
Normally a cold pin point gate would be constructed as shown below or into a depression which permits some gate vestige to still be below flush with top of part. The
spherical radius depression might be 0.020" - 0.030" deep into part (inside of part or
core side to follow same radius to maintain wall thickness).

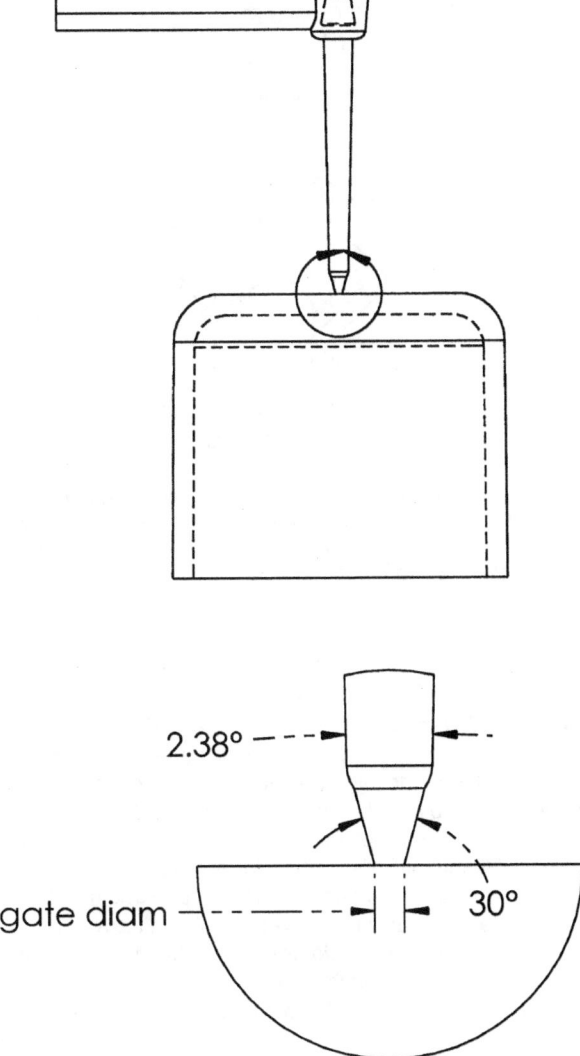

Runners (Cold)

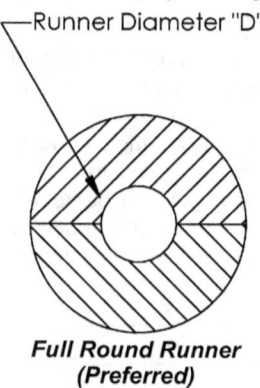

Runner Diameter "D"

Full Round Runner
(Preferred)

Modified Trapezoidal Runner

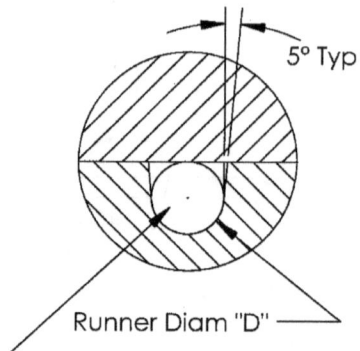

5° Typ

Runner Diam "D"

This is typical design for trapezoidal runner. A full round runner is best, but requires cutting runner in both mold halves. Not usually a problem, except when stripper plate is planned for ejecting runner such as three plate molds.

What makes full round best, and this design for trapezoid next best choice, is the ratio of CSA (cross sectional area) versus surface area. To accomplish lowest pressure drop we need largest CSA versus surface area. The surface area is where the heat transfer (cooling) will take place. We do not want premature cooling of the runner while fill and pack take place. Ideally there would be no cooling of runner until gate has sealed, and then fast cooling of both part and runner! This is not reality; thus, we choose runners which geometrically accomplish low pressure drop and slower cooling; this results in effective fill and pack. We may eject runners which are still soft and molten, but that is OK if we can get runner out of mold. We must take care to size runners as needed – smallest size that will fill and pack part. This is often smaller than designers might think. Mold designers often think the runner design requires a shape for faster cooling such as a true trapezoid with flat bottom and less depth, but the runner's first priority is effective fill and pack of the part.

Runners (cold - 3 plate)

Three plate mold, cold runners use trapezoid shape or modified trapezoid (round in bottom of channel) ... so runner can be stripped off sucker pins. The designs below are useful with softer olefin type resins which might collapse into sprue recess on three plate molds during strip action. The gusset in graphic below can be as shown or curved as an arc ... either way can supply added stiffness so that the runner does not collapse into stripper plate hole/recess for sprue bushing.

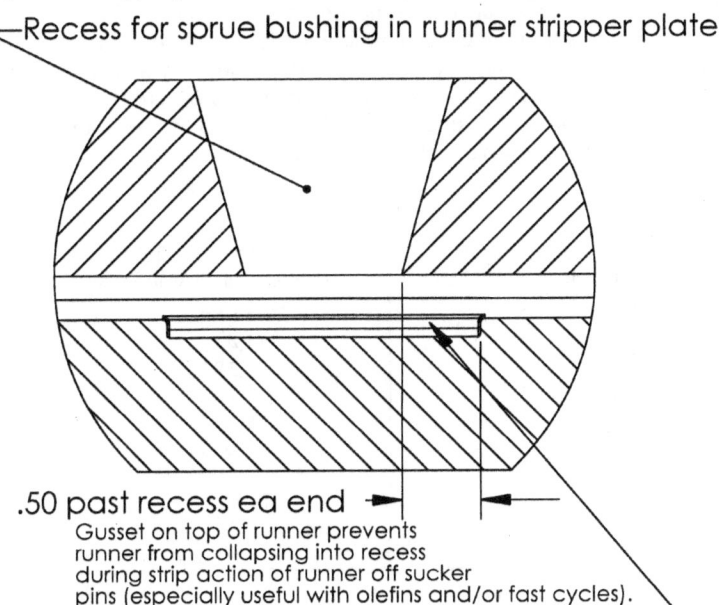

Recess for sprue bushing in runner stripper plate

.50 past recess ea end

Gusset on top of runner prevents runner from collapsing into recess during strip action of runner off sucker pins (especially useful with olefins and/or fast cycles). Draft each end or sweep out with radial groove on ends. Depth may need to be deeper than 40% of runner "D" when runner is very small.

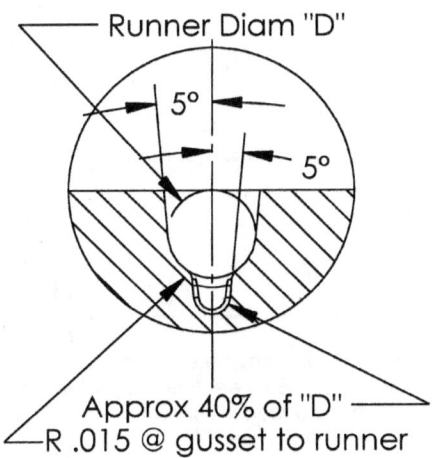

Runner Diam "D"

5°

5°

Approx 40% of "D"

R .015 @ gusset to runner

Scale 2X adjacent view

Page 43

Runner Sizing

Runner sizing should be done per the following:

A. Identify maximum part wall thickness.
B. Size runner by working from part back to sprue.
C. Set runner adjacent to gate at diameter equal to maximum part wall thickness.
D. Size each upstream branch by taking the number of branches (typ 2) to the 1/3 power X previous size. NOTE: $2^{1/3} = 1.259$
E. Repeat previous step until all branches sized.
F. If primary (largest) runner sizes appear too large and may lengthen the cycle, then oversized runners may be reduced in diameter by a selected percentage. A mold filling analysis can be done to identify fill pressures.
G. Remember to err on the small or "steel safe" side; it is easier to cut the runner bigger after the first mold trial than to resize smaller.

NUMBER OF BRANCH RUNNERS	RATIO OF FEEDER TO BRANCH DIAMETER
2	1.259
3	1.442
4	1.587
5	1.709
6	1.817
7	1.913
8	2.000

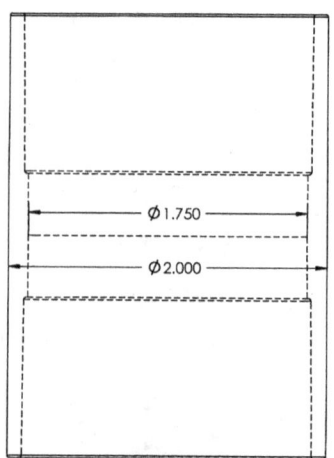

If MAX part wall is 0.125, then make last runner feeding part equal to part – 0.125 diameter. Size each next upstream runner using multiplier in table above based on number of branches – typically two, but could be more than two.

Runner Sizing Example
(balanced 16 cavity mold)

Location	Formula	Calculated runner size (inches)
max wall		0.125
A	equal to max wall	0.125
B	=1.259 x 0.125	0.157
C	=1.259 x 0.157	0.198
D	=1.259 x 0.198	0.249

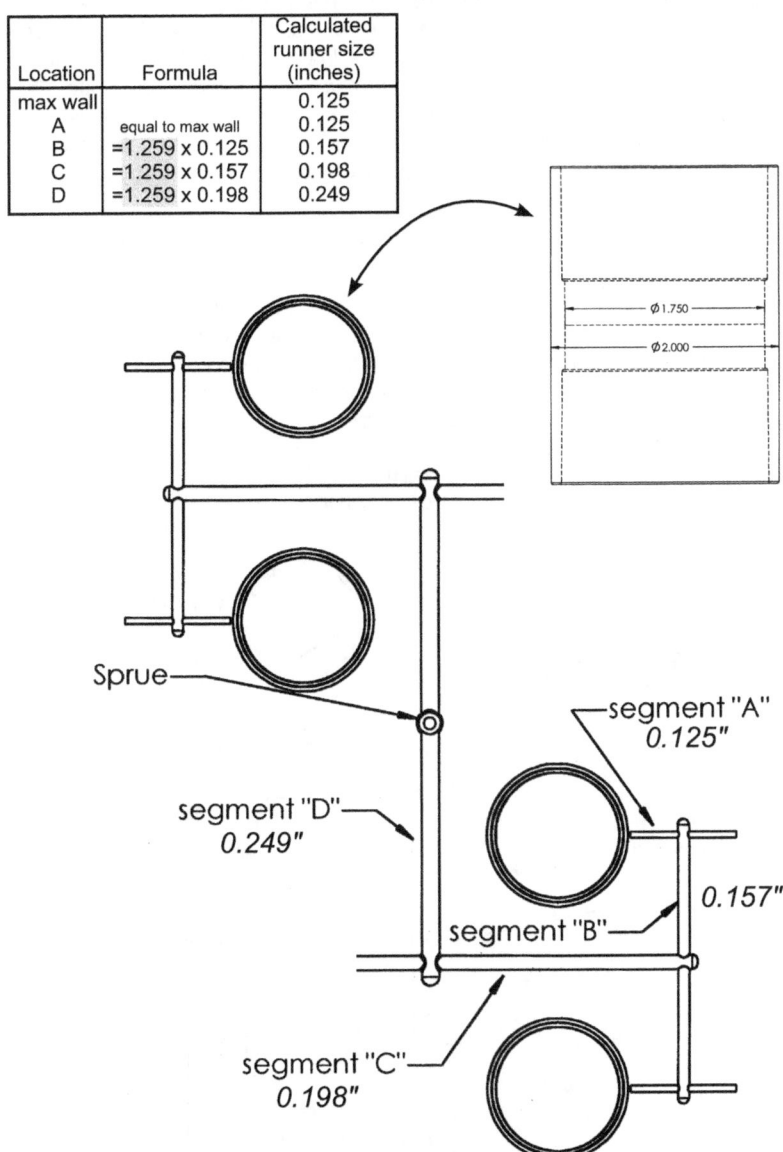

Sprue

segment "A"
0.125"

segment "D"
0.249"

0.157"

segment "B"

segment "C"
0.198"

NOTE: As stated above, multiplier is based on # of branches. If the secondary runner branch "C" fed an X pattern supplying four paths (4 branches direct to part); then there would be one less segment and/or size. The segment feeding final X pattern would need to be bigger since there would be 4 branches (0.125 X 1.587=0.198; same as current branch "C" since there would not be a branch currently known as "B").

Runners

In the runner example from preceding page, there are four levels of runner. The three branches each divide and turn 90°. Each intersection also has a cold well at end of feeder branch. With modern day CAD systems and CNC machining centers, runners can easily be made with sweeping bends. In the diagram below, it can be seen that the runner weight will be less for the top half design versus the bottom half design because the total length is less for some branches .....this results in reduced runner weight and reduced pressure drop in system. With regards to cold wells at each branch, the fast fill speeds often used today do not require as many cold wells beyond the one at base of sprue.

As mentioned earlier, always bias runner to smaller size if undecided as you can enlarge it easier than making it smaller.

This runner system is for a 16 cavity mold even though only four cavities are shown. Many molds such as this and larger today would use a hot runner system. Even if a cold, side located, edge gate is required for the part ... the mold could include a hot runner system. A four or eight drop HR system on this mold would greatly reduce runner scrap and often make post mold part handling more simple.

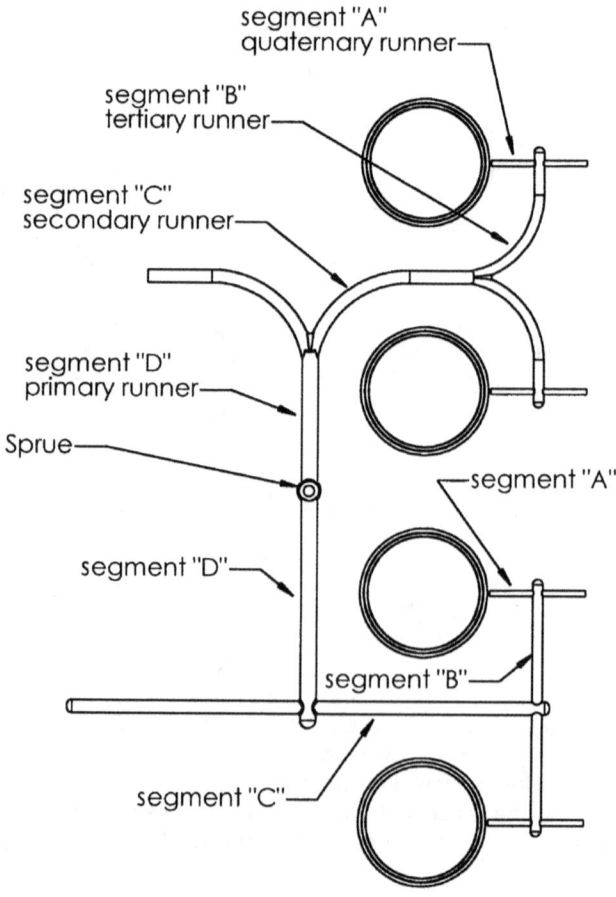

segment "A"
quaternary runner

segment "B"
tertiary runner

segment "C"
secondary runner

segment "D"
primary runner

Sprue

segment "A"

segment "D"

segment "B"

segment "C"

Sprue Pullers

Sprue pullers come in various designs. The example below shows a typical back-drafted puller or "coldwell". This serves two purposes: 1) sprue puller and 2) recess to accumulate cold slug of material from nozzle tip; thus, this puller is sometimes referred to as the coldwell. The adjacent ejector pin is sometimes referred to as the coldwell pin. With this type puller it is important to make the thru hole at parting line larger than ejector pin for two reasons: 1) the ejector pin must be able to pass thru; and 2) the extra 0.030" diameter plus the extra length 0.030" helps prevent the ejector pin from pinching plastic and shearing it off as ejector pin passes thru diameter at parting line. The extra 0.030" at face of ejector pin also insures that the ejector pin is not nested into plastic formed by coldwell. Alternate puller designs are also shown.

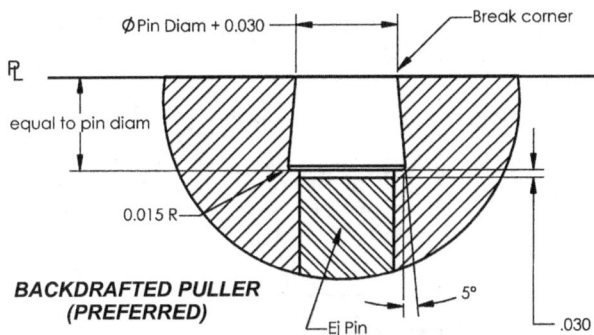

BACKDRAFTED PULLER (PREFERRED)

Intersecting runner not shown in this view; use 3° for gate pullers

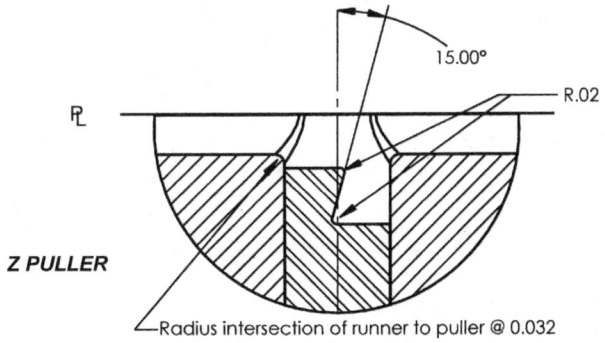

Z PULLER

Radius intersection of runner to puller @ 0.032

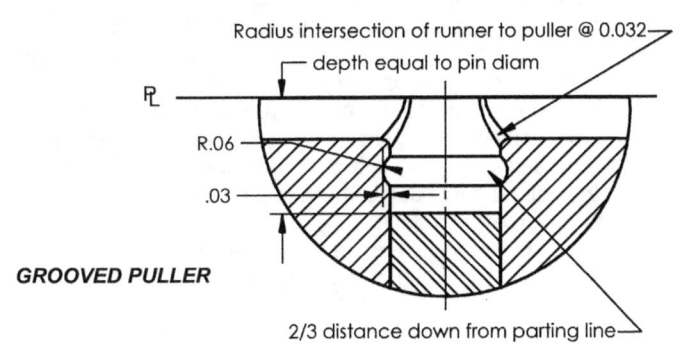

GROOVED PULLER

2/3 distance down from parting line

Sprue Sticking Solutions

Sticking sprues are often solved by any or all of the following (amorphous PET typically requires #7 below):

1. Increase nozzle temperature.
2. Reduce bubbles at base of sprue where it intersects with runner; bubbles can impede sprue pulling by the runner or coldwell; (use increased hold time or lower melt temperature; overpacking may make sprue tighter).
3. Check nozzle orifice ID versus sprue orifice ID: nozzle should be smaller (approx. 1/32" is typical).
4. Check and correct nozzle alignment with locating ring.
5. Remove scratches and grooves caused by steel tools inserted into sprue bushing ID. To correct: Recut with tapered reamer (done by machine shop or tool room).
6. Use backdrafted coldwell or other suitable puller design such as types shown at right.
7. If sprue is large requiring excessive cooling time; the resin must be cooled sufficiently to release from sprue ID. To correct: Make sure sprue bushing is in contact with moldbase steel for full length; eliminate air gaps; improved cooling results from a line on line fit of bushing to hole. Sprue bushings are available in high conductivity copper alloy which helps with faster sprue cooling (better bushing thermal conductivity).

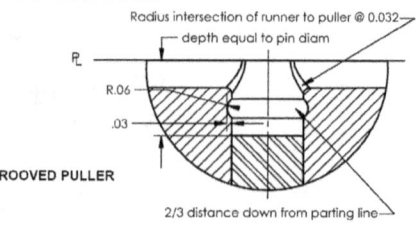

8. Increase cycle time to accomplish more cooling.
9. Some resins which have poor lubricity and require high packing may benefit from a sprue taper of 3/4 inch per foot (3.57° incl angle) instead of the standard 1/2 inch per foot taper (2.38° incl angle). Do # 7 first.
10. Poor nozzle spherical radius match to sprue bushing's spherical radius. Recut same spherical radius on each (done by machine shop or tool room).
11. Use a nucleating agent which accelerates the setup of crystalline resins. Colored resins typically setup (solidify) faster than do clear or natural resins since the colorant acts somewhat like a nucleating agent.
12. Draw polish sprue ID.
13. Use an electrically heated hot sprue bushing.
14. Try water cooled sprue bushing from HASCO®.

Sucker Pins

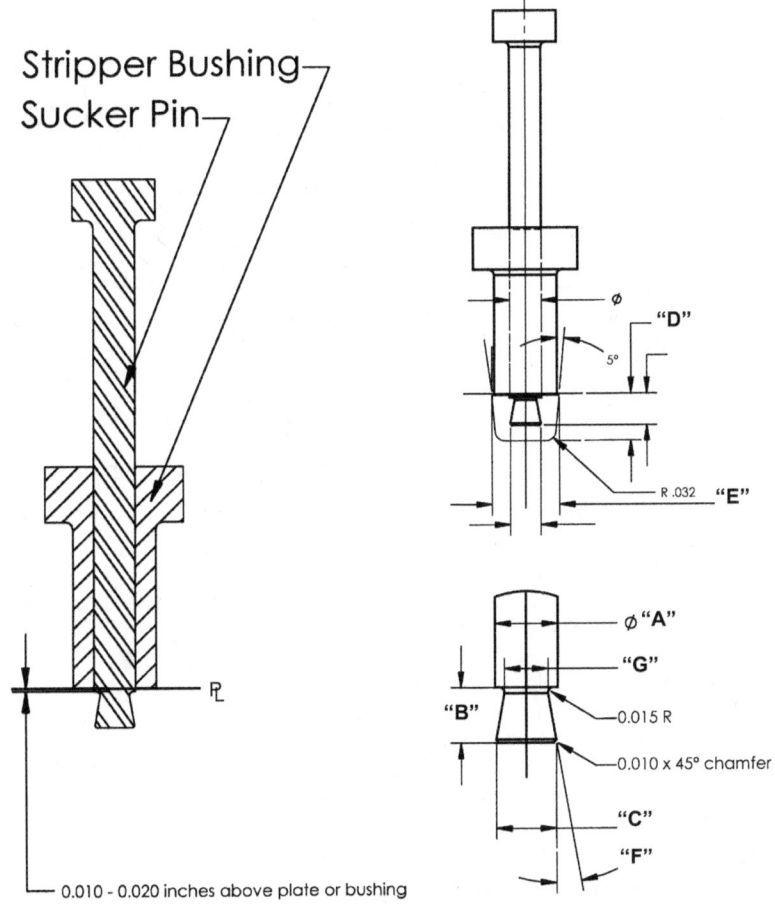

Stripper Bushing
Sucker Pin

"D"
5°
R .032 "E"

⌀ "A"
"G"
"B"
0.015 R
0.010 x 45° chamfer
"C"
"F"

0.010 - 0.020 inches above plate or bushing

		1/8" Pin	3/16" Pin	1/4" Pin	5/16" Pin
A	Pin Diam	0.125	0.188	0.250	0.312
B	Undercut Length[1]	0.110	0.165	0.220	0.275
C	Undercut Diam	0.120	0.180	0.240	0.300
D	Pin Well Depth	0.169	0.253	0.338	0.421
E	Pin Well Diam	0.250	0.325	0.403	0.450
F	Undercut Angle[2]	5 -10°	5 -10°	5 -10°	5 -10°
G	Undercut Small Diam	0.089	0.133	0.178	0.222

[1] Undercut length includes radius and chamfer; effective length is less.
[2] Use 5° for PS, SAN, PMMA; use 7½° for POM, PA, PET, PC, HIPS, ABS; use 10° for PVC, PP & PE; whenever greater pull is required, use larger pin to permit greater angle so as to avoid "G" dimension smaller than 2/3 x pin diam for adequate strength. When in doubt of angle: use 5°... can then recut to larger angle if needed. Very stiff resins such as PPS and LCP may need to start at 3°.

Venting

NOTES:
1. Vent land is typically 0.04" – 0.12" long; 0.090" suggested for perimeter vents (see page 52 on perimeter venting).
2. Vent relief is 0.03" deep & 0.25" – 0.5" wide ... vent relief must run out to atmosphere.
3. Vacuum venting[1] should only vent to vacuum source
4. Merits of vacuum venting[1] are:
a. Longer interval between cleaning (volatiles pulled farther up the vent relief channel).
b. Can accomplish same venting with vent of reduced depth; important when vents must be very shallow.
5. Vents should be ground in.
6. Vents & relief should be consistent throughout common mold components.
7. Deeper vents can be used with GF resins.
8. Cold runner system should also be vented.
9. Important ... High melt flow rate resins require vents of lesser depth.
10. Vents that are in a moving pin are self cleaning and preferred over non-moving pins.
11. Make vents as accessible as possible for cleaning purposes.

These are general guidelines; refer to supplier design manuals for additional information.

---

[1] Vacuum supplied from a purchased vacuum generator which typically includes some form of a jet pump – high pressure air is converted by the nozzle to create high velocity which creates suction. Penberthy is a manufacturer of jet pumps; see p 103 for possible address.

Vent Depths for Plastic Resins

VENT DEPTHS FOR VARIOUS RESINS	
RESIN	DEPTH (inches)
ABS	0.0010 - 0.0015
ACETAL	0.0005 - 0.0010
ACRYLIC	0.0015 - 0.0020
CELLULOSE ACET, CAB	0.0010 - 0.0015
ETHYLENE VINYL ACET.	0.0010 - 0.0015
IONOMER	0.0005 - 0.0010
LCP	0.0005 - 0.0007
NYLON	0.0003 - 0.0005
PPO/PS (NORYL)	0.0010 - 0.0020
POLYCARBONATE	0.0015 - 0.0025
PET, PBT, POLYESTERS	0.0005 - 0.0007
POLYSULFONE	0.0010 - 0.0020
POLYETHYLENE	0.0005 - 0.0012
POLYPROPYLENE	0.0005 - 0.0012
POLYSTYRENE	0.0007 - 0.0010
POLYSTYRENE (IMPACT)	0.0008 - 0.0012
PVC (RIGID)	0.0006 - 0.0010
PVC (FLEXIBLE)	0.0005 - 0.0007
POLYURETHANE	0.0004 - 0.0008
SAN	0.0010 - 0.0015
T/P ELASTOMER	0.0005 - 0.0007

Perimeter Venting

With today's fast fill times, venting the gases from the cavity can become more of a challenge. The figure below shows a cavity with venting (parting line vents) around the full perimeter of the cavity. This cavity is center gated, but corners will still be last place to fill unless wall thickness is adjusted slightly to balance (or flow leaders added to bias flow into corners to keep up with slightly shorter flow length to sides). Many tooling engineers like to make full perimeter vents slightly less deep than they would localized vents to safeguard against flash – can grind in deeper later (easier than reducing depth). Note: If it does not flash as a localized vent, it should not flash at same depth for full perimeter vent. Full perimeter vents need a full perimeter vent relief as well, to collect the vented gases; dumps to atmosphere can obviously be localized as there must be some projected steel to support clamp tonnage without hobbing the parting line.

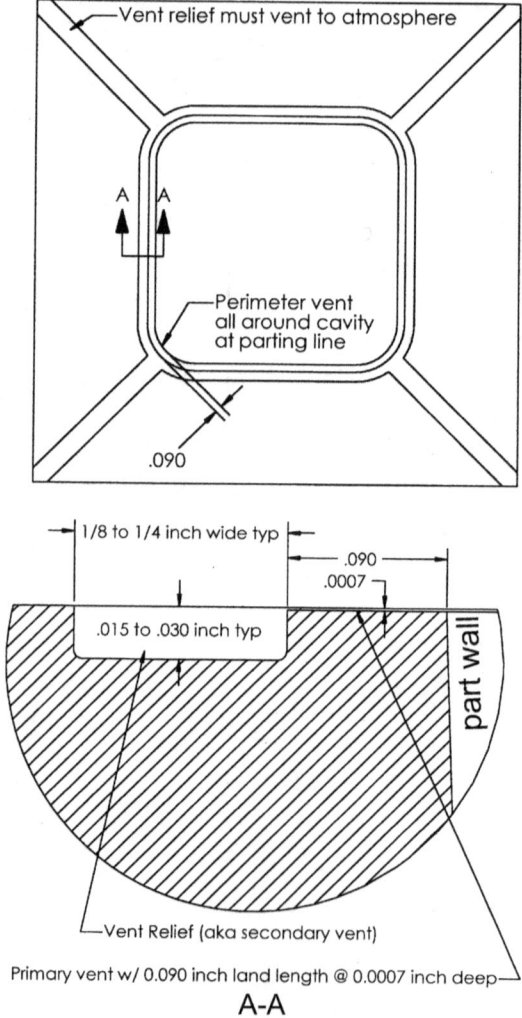

A-A

Blank Page

Ejection

Ejector System Requirements

1. The mold should eject part automatically without part distortion.
2. The ejector plate should be guided on a minimum of four hardened steel pins sliding in self lubricated bronze bushings (bushings with grease grooves if self lube not available). The support pillars and/or return pins do not provide sufficient alignment by themselves.
3. The plate that locates the alignment pins must be water cooled or thermally isolated from the cores by a water cooled plate to prevent the ejector plate from binding on the pins due to thermal expansion.
4. Limit forward travel of ejector plate to that required to eject parts.
5. Ejector plate should have K.O. extensions as a standard unless otherwise specified. K.O. extensions should extend to be 0.010" <u>below flush</u> with clamping plate. K.O. extensions (see figures below) can be threaded into ejector plate or have head with flat located in counterbore in ejector plate. The ejector plate (and KO extensions) should be tapped with 5/8-11 threaded holes (1/2-13 on smaller molds) at K.O. locations to enable the K.O.s to be attached to ejector plate making positive pullback possible.

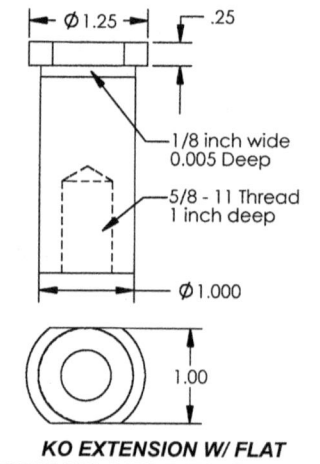

KO EXTENSION W/ FLAT
RECESSED INTO EJECTOR PLATE²

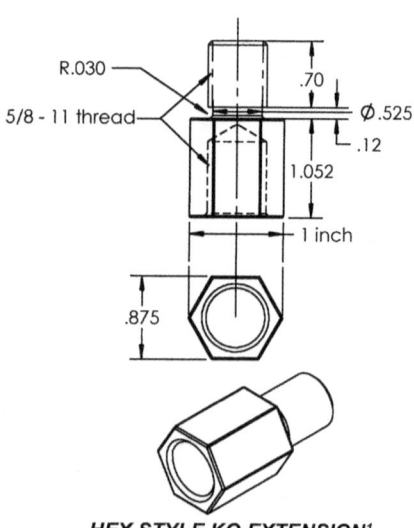

HEX STYLE KO EXTENSION¹

Note:
Socket diameter approx 1.185" – will work well with 1¼" thru hole in bottom clamping plate.

---

¹ Length includes 0.875" for clamp plate + 0.187" for stop button minus 0.010"

² Length dependent on clamp plate thickness; use of stop buttons and thickness of ejector plate. Recess head into ejector plate, and <u>not retainer plate</u>. Locate SHCS adjacent to KO extension holding retainer plate tightly to ejector plate as KO extension now pushes directly on retainer plate.

Ejection (6-13 ... early return)

6. Positive pullbacks are preferred over return springs due to increased reliability – springs break.
7. Thru hard ejector pins are preferred (e.g. DME "THX" has core hardness of 50-55 Rc and a surface hardness of 65-74 Rc; head is annealed to 30-35 Rc.
8. Ejector pins should have bearing surface equal to at least two diameters, but not more than one inch in length ... = 2d < 1".
9. When the ejectors pins (or lifters) are located such that they can contact the cavity (opposite mold half) during mold closing, the ejector plate return pins should be spring loaded to prevent this from happening (e.g. a mold with "thumbnail" style ejector pins); see figure below. Companies such as DME and HASCO do sell devices which accomplish same thing via mechanical devices with action which is more positive having less chance of malfunction - typically listed as "ejection early return devices".

Pin Diam	"A" Diam	"B" Diam
0.500	0.52	0.77
0.625	0.64	0.89
0.750	0.77	1.02

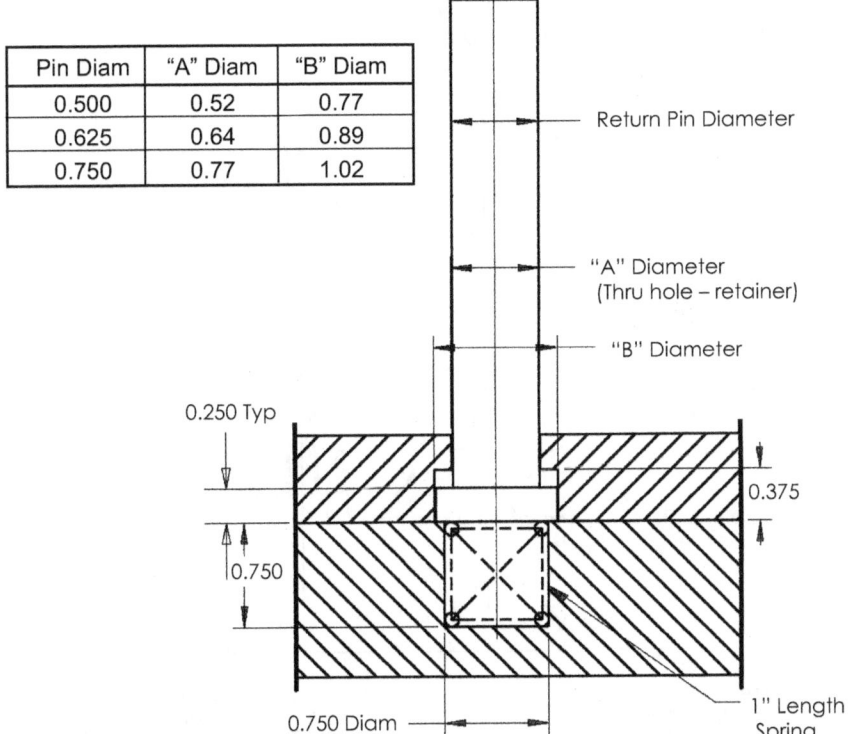

10. When accelerated ejector pins are used, some provision must be made to mechanically return the ejector plate the additional travel distance without the pins hitting the "A" half of the mold.
11. Contoured ejector pins located to prevent rotation (flat is preferred; see next page).
12. Ejector system should move freely when mold is in the operating position.
13. Ejector plate should have positive stop buttons in line with return pins.

Ejection (keyed ejector pins)

Ejector pins which are contoured to match shape of part (e.g. parabolic reflector molds for lamp hsgs, etc) must have the ejector pin keyed in a manner such to prevent turning. As can be seen below, if the pin on right was turned 180° it would likely core into molded wall thickness and hit cavity causing mold damage and/or a molded part defect. These contoured or profiled pins are keyed via one of several methods (either of first two preferred):

1. Single flat and enlarged recess (larger than double flat shown below with different center line to pin head's center). Offers the advantage of only one way to install – the correct way. See also design guidelines on next page.
2. Double flat as shown below with narrow, but elongated recess; some say this provides best location, but allows pin to be loaded 180° from proper location.
3. Flat with adjacent groove and parallel dowel pin or key laid in to locate (not shown ... requires added part which can be left out).
4. Drilled hole in head with small dowel pin perpendicular to head extending into adjacent milled recess (not shown).

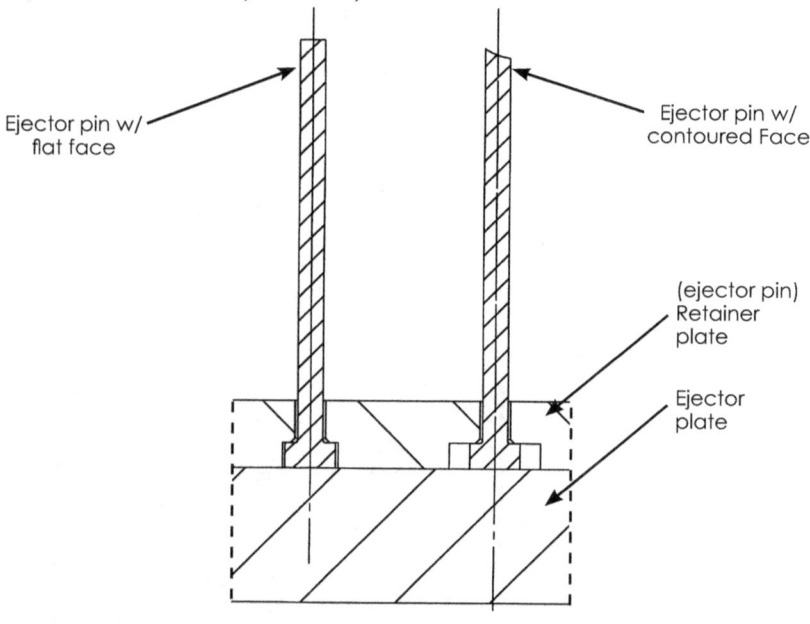

Ejector pin w/ flat face

Ejector pin w/ contoured Face

(ejector pin) Retainer plate

Ejector plate

Flat faced pin can use full round counterbored recess for pin head in ejector pin retainer plate. Note the chamfer for ejector pin to head radius; this radius on ejector pins may be up to 1/32 inches. A chamfer helps with installation and provides clearance for radius on pin

Single Flat to orient contoured ejector pin. Note: Pin cannot be installed incorrectly.

Double flat also orients contoured pin; may provide better location and uses less space. Note: Pin can be loaded 180° from correct orientation unless flatted on three sides to prevent.

Ejection (keyed ejector pin design table)

Keyed Ejector Pin Design Guidelines
(continued from previous page)

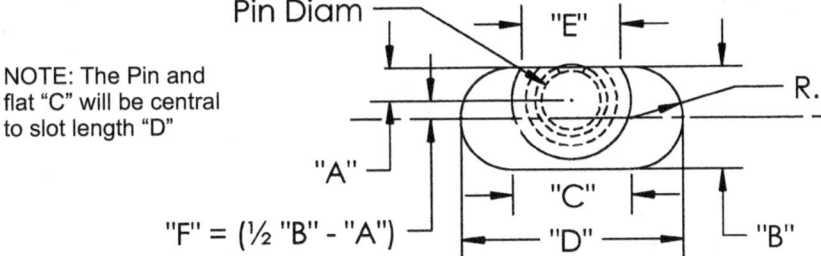

Pin Diam

NOTE: The Pin and
flat "C" will be central
to slot length "D"

"A"

"F" = (½ "B" - "A")

"E"

R.

"C"

"D"

"B"

Pin DIAM (IN)	Flat To Pin Center "A" (X.XXX) (IN)	Head Diam & Slot Width "B" (X.XX) (IN)	Slot Flat Length "C" (X.XX) (IN)	OA Slot Length "D" (X.XX) (IN)	Pin Flat Length "E" (X.XXX) (IN)	Hole Center to Slot Center Line "F" (X.XXX) (IN)
1/8	0.080	0.25	0.25	0.50	0.192	0.045
5/32	0.092	0.28	0.27	0.56	0.213	0.049
3/16	0.125	0.38	0.34	0.72	0.280	0.063
7/32	0.138	0.41	0.36	0.77	0.298	0.065
1/4	0.160	0.44	0.36	0.80	0.298	0.059
9/32	0.160	0.44	0.36	0.80	0.298	0.059
5/16	0.200	0.50	0.36	0.86	0.300	0.050
3/8	0.250	0.63	0.44	1.06	0.375	0.063
7/16	0.275	0.69	0.47	1.16	0.413	0.069
1/2	0.300	0.75	0.51	1.26	0.450	0.075
9/16	0.325	0.81	0.55	1.36	0.488	0.081
5/8	0.350	0.88	0.59	1.46	0.525	0.088
3/4	0.400	1.00	0.66	1.66	0.600	0.100
7/8	0.500	1.13	0.58	1.70	0.515	0.063
1	0.550	1.25	0.66	1.91	0.594	0.075

Air Poppets

Air Poppets

During ejection, some part characteristics require air poppets to supply enough air to avoid generating a vacuum between part and mold steel. Generous venting may not supply sufficient air to avoid push marks as ejector system pushes against the vacuum generated. The ejector system is strong enough, but often the parts are not! Such problems are common when molding flexible houseware items made from PE & PP. There are various suppliers of air poppet valves such as the one shown below from PCS Company. Note: There will be two witness lines on the part (poppet body + valve). Note: May not want to use water fittings as shown in drawing below ... use suitable fitting such that mold setter cannot make mistake and connect coolant to air circuits.

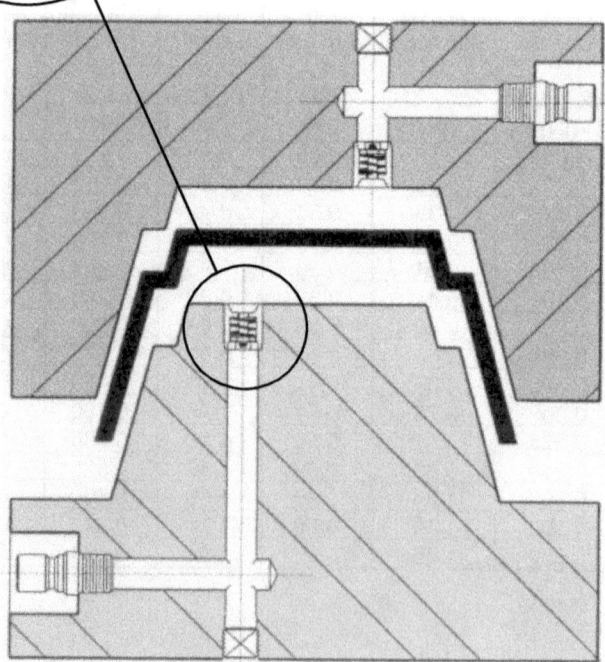

Figures;
courtesy of:

PCS Company
34488 Doreka Dr
Fraser, MI 48026
(800) 521-0546

Ejection (ejector & retainer plate design)

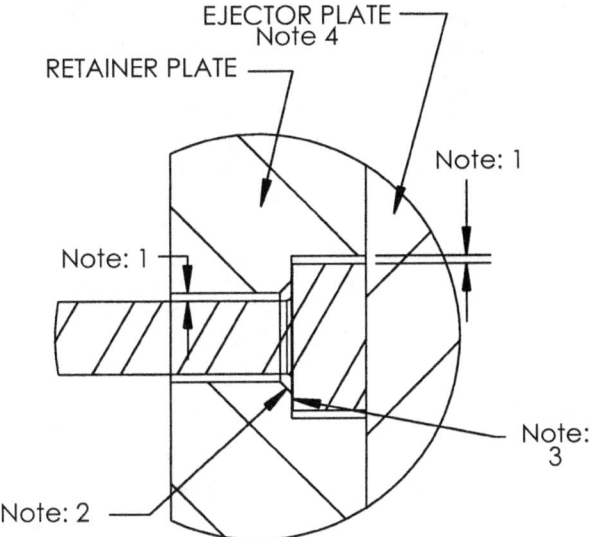

NOTES
1. Provide 0.010" - 0.020" clearance in retainer thru hole and counterbore to help pin self align with hole in core where bearing surface is located. On very hot molds, the thermal expansion should be calculated for pin farthest from mold center and compared to this clearance to insure the pin is not in a bind from core expanding away from mold center. When this movement is large; the ejector plate can be drilled to include coolant lines, but guide pin for ejector system should be located in similarly heated plate.
2. Ejector pins can have up to a 1/32" radius where shaft meets head (radius is often less). A 45° x 0.032" chamfer will provide clearance for radius and help ease mold assembly.
3. The counterbore depth should provide minimal clearance for pin head height at 0.001" - 0.003".
4. The ejector plate is thicker than retainer as it must transfer sufficient ejector force without warping under load. If and when a single center KO bar is used, then a thicker ejector plate must be used to withstand force without warping (always use guide pins and bushings). Multiple KO bars are suggested which incorporate the largest pattern possible (i.e. 4" x 16" pattern preferred over 7" pattern).
5. Ejector pin length (overall) – Some molders want the pin to be long so as to mold a recessed impression in part so no projections interfere with post molding assembly equipment. Other molders want the pin to be short so as to mold a small projection to insure parts fall with no chance of pin being nested into part whereby part is stuck or molded onto pin. Check with customer whether molded pin impressions are flush or below vs flush or above in molded plastic.

Ejector System (guided)

The figure below shows a guided ejector system whereby the ejector plate is guided by steel pins and bronze bushings. Also listed are various other components pertinent to the ejector system. Shoulder type leader pins enable the moldmaker to line bore both leader pin and bushing holes to accomplish optimum alignment.

The guide pins for the ejector system may be located in the support plate as shown OR located in the bottom clamping plate. In the view shown here whereby guide is located in support plate, if the cores are run hot and support plate was running at 190° F, then the thermal expansion would move the guide pins outward causing system to bind. In that event, the guide pins would be better located in bottom clamping plate OR add coolant lines to ejector plate (two running long axis of mold are often sufficient unless mold is very large). If the mold runs hot and you have decided to locate guide pins in bottom clamping plate and have elected not to add coolant lines to ejector plate (for heating) ... you must calculate thermal expansion and compare to ejector pin clearances in retainer plate.

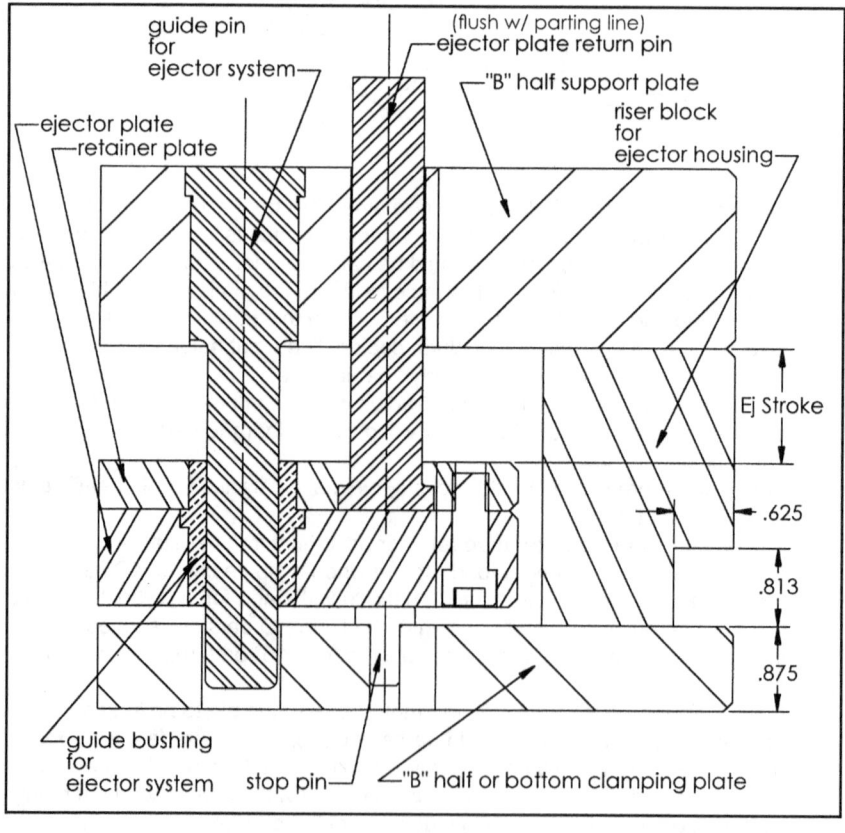

Ejector System (delayed pins)

Delayed Ejection

Delayed ejection can be accomplished by adding a second pin inverted so head supports head. After ejector plate travels counterbored depth creating delay, the ejector plate will push both pins forward. This can be used to delay the ejection of runner adjacent to gates so as to create in mold degating of some smaller chisel type edge gates. Timed, multiple plates are preferred when there are many pins to delay or more complex actions, sleeves, etc.

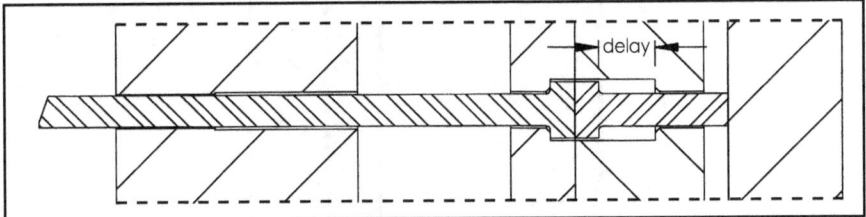

Ejection (tied to stripper plate)

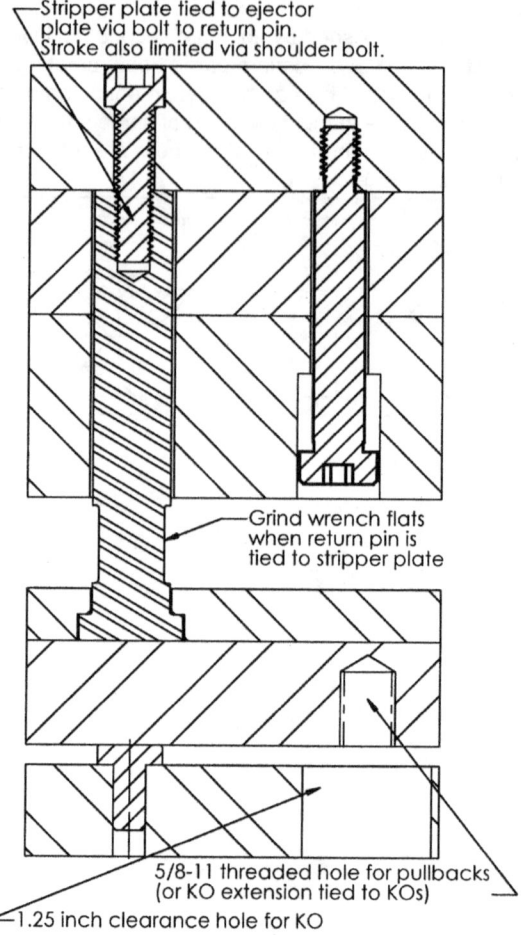

Stripper plate tied to ejector plate via bolt to return pin. Stroke also limited via shoulder bolt.

Grind wrench flats when return pin is tied to stripper plate

5/8-11 threaded hole for pullbacks (or KO extension tied to KOs)

1.25 inch clearance hole for KO

In example above, the ejector plate is attached to a stripper plate ejection system via return pins. The ejector stroke is limited by the shoulder bolt as shown.

The single clearance hole for the KO extension is shown out of position for this section view for viewing purposes. Ejector plate pullbacks are preferred to springs as the springs will eventually break. If springs are used, take care to minimize the compressed height – this requires close attention to spring supplier recommendations and using longer springs as best spring life comes from limiting compression to about ¼ of free length which may only be 15 - 20% of usable stroke since there is also preload required which already uses some length.

Ejection (SPI KO patterns)

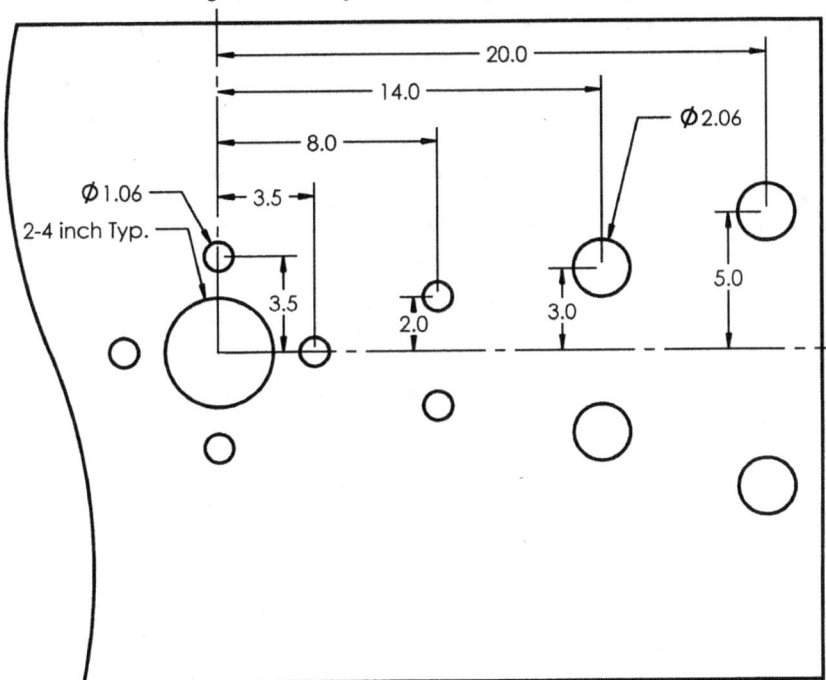

Note: The next larger pattern (not shown) in diagram above is 16" x 52" (very large press).

A 300 ton press will likely have only the center, 7" H & V + 4 x 16 pattern (some would call the 4 x 16, a 2 x 8). Some presses have these patterns as both horizontal & vertical, but some presses may have certain patterns missing based on platen size and other press features which may interfere (e.g. linkage attachment on toggle press). Hydraulic clamp presses often have more KO holes from the possible SPI patterns. Use largest possible pattern available on press.

---

[1] Remember that not all hole patterns are available in all presses. A 110 ton press may have a 4 x 16 pattern horizontally, but not vertically. A press from Japan may have JIS KO pattern which is slightly different. A press from Europe may have Euromap pattern which is slightly different. Always request platen drawing for end use molding press before mold design. In stripper example on previous page, if shoulder bolt reached across parting line to "A" half of mold, then stripper would be actuated by mold opening – some toggle presses may have a minimum opening stroke!

Ejection (vented ejector pins)

The graphic below shows typical ejector pin land lengths as well as typical geometry for vented ejector pins. This same geometry can be placed on a non-moving pin or blade to create a vent, but locating the vent on an ejector pin makes it a "self cleaning" vent. Some amount of self cleaning is inherent with the pin movement; the pins can also be manually held forward for added cleaning. Round pins with flats each 90° do not result in effective venting as the vent depth is only at nominal depth in four places which are lines of negligible width (see figure below); thus, a full circumferential vent is best (or full width as installed on a non-round, blade insert).

Vented Ejector Pins (full round vs flats)
Vented Ejector Pins (full round vs. flats)

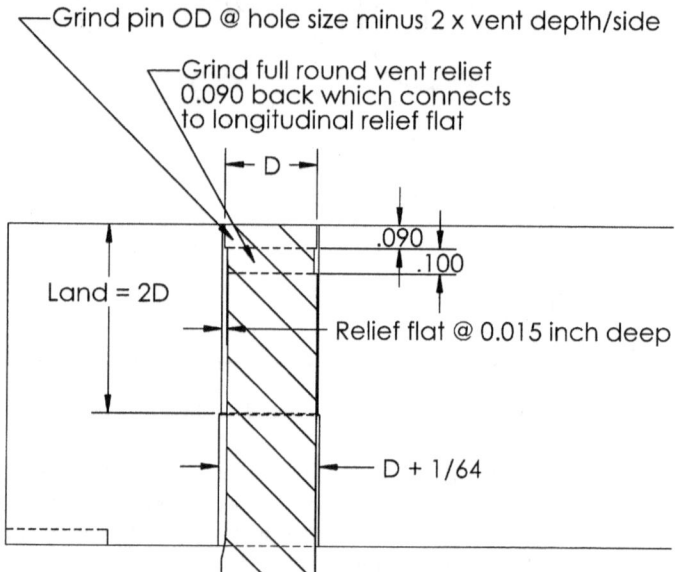

Grind pin OD @ hole size minus 2 x vent depth/side

Grind full round vent relief 0.090 back which connects to longitudinal relief flat

D

.090
.100

Land = 2D

Relief flat @ 0.015 inch deep

D + 1/64

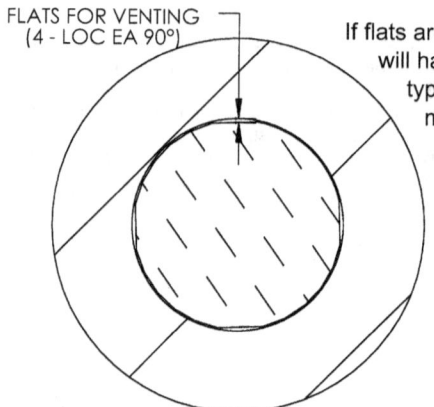

FLATS FOR VENTING
(4 - LOC EA 90°)

If flats are used to create vents on a pin, the flats will have to be deeper than the 0.0005" - 0.002" typical vent depth since that depth is only at middle of flat length (maybe as much as 5-10X greater than normal). Since this is hard to predict as to what depth is effective without flashing, then the full round vent is preferred AND located in a moving pin to make it "self cleaning".

Hydraulic Cylinders

CYLINDERS, HYDRAULIC AND PNEUMATIC

1. Limit switches should be present to control hydraulic and pneumatic cylinders.
2. Hydraulic or pneumatic cylinders used for plate movement should be connected to provide one extend and one retract fitting.
3. Self-contained hydraulic ejector/core pull systems should be equipped with quick disconnect couplings between cylinders and flexible hose. Use Parker H6-63111 for male fittings and Parker H6-62111 for female fittings (3/4" fitting). The type of fitting might be as follows:

	EJECTORS	CORE PULLS
Mold:	Extend – male quick disconnect	Set – male quick disconnect
	Retract – female quick disconnect	Pull – female quick disconnect
Machine:	Extend – female quick disconnect	Set – female quick disconnect
	Retract – male quick disconnect	Pull – male quick disconnect

4. Design mold with mechanical stops in both stroke directions.
5. All hydraulic cylinders shall be rated for a minimum of 3000 psi, but sized for 1500 psi operating pressure (e.g. Parker 2H heavy duty hydraulic cylinder). A typical cylinder may be as follows:

Vertical labels (left to right): BORE DIAMETER · CUSHION ON HEAD END · RECTANGULAR FLANGE MTG ON HEAD · 2H SERIES · SAE STRAIGHT THRD O-RING PORT · STD ROD DIAMETER · STD ROD END (SMALL MALE THRD) · UNF STD THRDS ON ROD END · CUSHION ON CAP END · STROKE LENGTH

2½ C J -2H T 1 4 A C-18

PARKER SH CYLINDER W/ TYPE J MTG FLANGE

TABLE OF VARIOUS COMMON CYLINDER DIMENSIONS

Bore	Rod Diam	Rod End Thread	Flange Footprint	Flange Thickness	Flange Hole Pattern	Flange Thru Hole
1½	0.625	7/16-20	2.5 x 4.25	3/8	1.63 X 3.438	7/16
2	1.000	3/4-16	3 x 5.125	5/8	2.05 X 4.125	9/16
2½	1.000	3/4-16	3.5 x 5.625	5/8	2.55 X 4.625	9/16
3¼	1.375	1-14	4.5 x 7.125	3/4	3.25 X 5.875	11/16
4	1.750	1¼-12	5 x 7.625	7/8	3.82 X 6.375	11/16

(Note: All Dimensions Are In Inches)

COMMON CORE PULL SEQUENCES

"A"	"B"	"C"	"D"	"E"
Core In/Set	Clmp Close	Core In/Set	Clamp Open	Core In/Set
Clmp Close	Core In/Set	Clamp Close	Eject	Clamp Close
Inject	Inject	Inject	Core Out/Pull	Inject
Clamp Open	Core Out/Pull	Clamp Open to	Core In/Set	Clamp Open
Core Out/Pull	Clamp Open	Interim Stop	Eject Retract	w/ Core Pull
Eject	Eject	Core Out	Clamp Close	Eject
Eject Retract	Eject Retract	Clamp Open	Inject	Eject Retract
		Eject		
		Eject Retract		

Slides (mechanical)

In this example, the slide has tapped holes in face for a bolt on slide core retainer (retainer and core not shown, but required pull is ½").

Design Guidelines:

1. Due to delay in pulling slide (action is a function of mold opening) the slide will be mounted to ejector mold half ....do not eject until slide is pulled. If ejector pins are located below slide, then limit switches are suggested to verify slide pull prior to ejection.
2. Wedge locking faces should be hardened surfaces.
3. The slide should be held in open position by detents, springs, etc. When stroke is reasonably short a flat can be added to horn pin and slide to make it self re-aligning ... slides get out of position during cycle interrupts for cleaning, part removal, etc. Depending on mold open stroke & space available in mold; the horn pin can be long enough so as to never disengage the slide.
4. Angle of the horn pin @ 15 - 25° ... do not exceed 25° & do not use less than 7°, 15 - 20° is optimum.
5. The angle on wedge or heel block should be at least 3-5° greater than horn pin angle.
6. Clearance hole @ 0.030 - 0.050" oversize.
7. Calculate required <u>working</u> pin length[1] as follows: (see also formula in graphic on next page):

    length = (req'd pull + hole clearance) x Tan (90°- pin angle)

---

[1] Not overall length as it does not include head or buried length or angle on end (or spherical radius if used).

Mechanical Slides (continued)

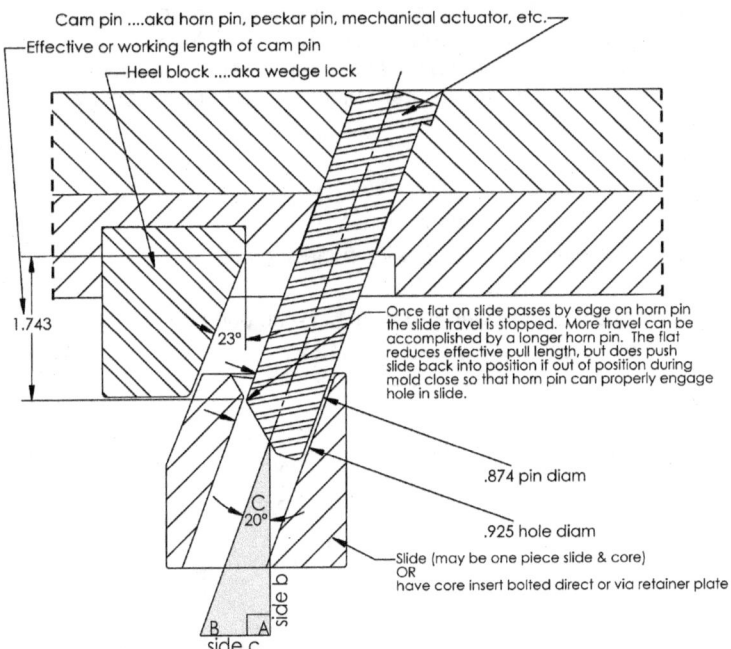

Cam pin ....aka horn pin, peckar pin, mechanical actuator, etc.

Effective or working length of cam pin

Heel block ....aka wedge lock

1.743

23°

Once flat on slide passes by edge on horn pin the slide travel is stopped. More travel can be accomplished by a longer horn pin. The flat reduces effective pull length, but does push slide back into position if out of position during mold close so that horn pin can properly engage hole in slide.

.874 pin diam

.925 hole diam

C 20°

Slide (may be one piece slide & core)
OR
have core insert bolted direct or via retainer plate

side b

side c

B   A

Required pull equal ½ inch
side b = side c x Tan B
side b = (pull + hole clearance) x Tan B
side b = (0.500 + 0.051) x Tan 70°
side b = 0.551 x 2.74747
side b = 1.514 min eff length of horn pin
Note : We have 1.743 which yields 0.083" extra pull

side c = (1.743 - 1.514) x Tan C
side c = side b x Tan C
side c = 0.229 x 0.36397
side c = 0.083 inches

Support Pillars

Support pillars are needed to provide additional support to strength of the support plate; without the pillars the support plate would require added thickness to avoid deflection. Deflection would create gaps at parting line and result in parting line flash (and possible binding of mechanical actions). The formulas below are used to calculate deflection.

Cavity plate or surface deflection

d_{max} = maximum deflection of plate in inches

F = total force of injected plastic (lbs) ... projected area X inj psi

w = width between supports in inches

E = modulus of elasticity; steel = 30 x 10⁶ psi = 30,000,000 psi

I = moment of inertia of cavity plate – Lt³/12 where L is cavity plate length and t is cavity plate thickness) in inches

$$d_{max} = \frac{Fw^3}{48EI} \qquad I = \frac{Lt^3}{12}$$

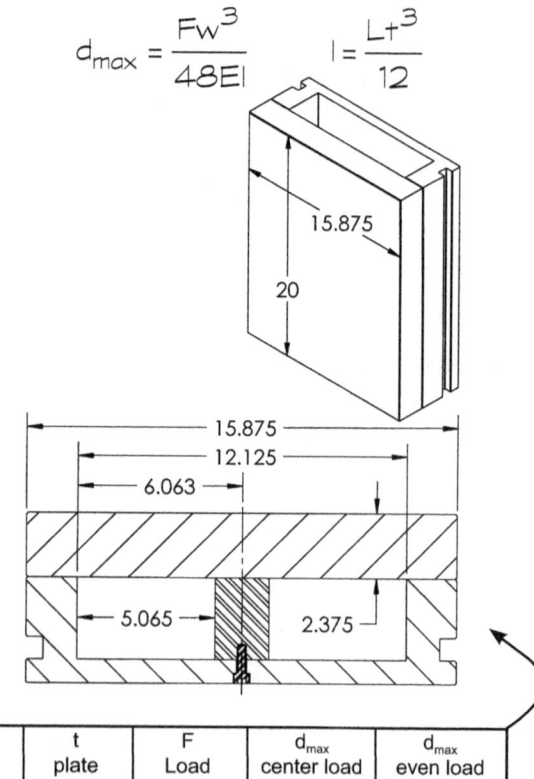

Note:
This formula is one for concentrated load (single point of contact in middle of span). When load is spread uniformly over more of span (or all of span), this formula would change the 48 to be 76.8 yielding a smaller deflection for same load. In calculations below right the projected area is 32 in² with 10,000 psi inj pressure.

w span width (in)	L plate length (in)	t plate thickness (in)	F Load or Force (lbs)	d_{max} center load deflection (in)	d_{max} even load deflection (in)
12.125	20	2.375	320,000	0.0177[1]	0.0111[1]
5.065	20	2.375	320,000	0.0013[2]	0.0008[2]

[1,2] footnotes at bottom page 69

Support Pillars (continued)

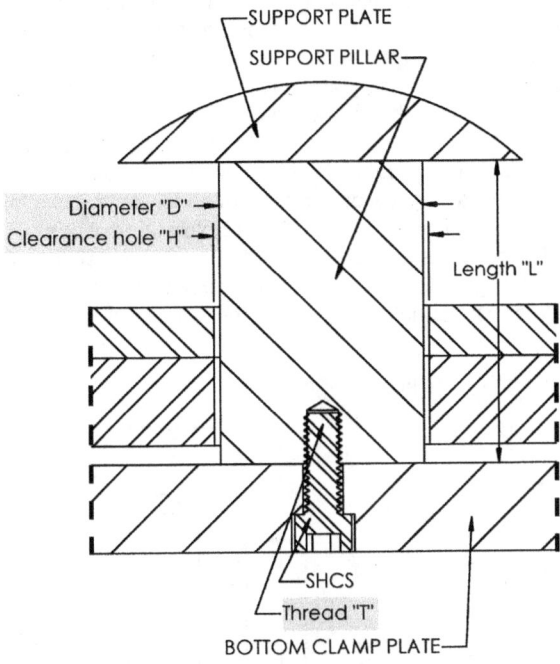

Support Pillars[3] ODs, Clearances & Threads		
"D" (in)	"H" (in)	"T" (in)
1.00	1.12	3/8 –16
1.25	1.38	3/8 –16
1.50	1.62	3/8 –16
1.75	1.87	3/8 –16
2.00	2.12	3/8 –16
2.50	2.62	3/8 –16
3.00	3.12	5/8 –11
4.00	4.12	5/8 –11

[1] If no support pillars.....obviously too much deflection (less than 0.001" suggested).

[2] Deflection for center row of supports; less deflection w/ two rows – 5.065 would be less; at a w = 3"; the deflection would be approx. 0.0003".

[3] Use 1040 CRS for #2 steel mold bases; suggest 630 SS 17Cr-4 Ni for SS bases.

Alignment

There are typically various forms of alignment in the mold all working together. They start with the locating ring which locates mold in center of platen. This permits alignment of mold's sprue bushing to the machine's injection nozzle (and sets up for proper KO rod location in clamping plate). There are leader pins on the mold to align one mold half with the other, but there are often additional locks (either tapered or straight) in the parting line(s). Note that since the mold is located in press by the mold's locating ring which is in the center of the mold, thermal expansion will occur about this center line. This is why the optimum location for taper locks (or straight locks) is at the centers of the four sides (3, 6, 9 & 12 o'clock positions). These locations are unaffected by thermal expansion about the mold's center line. See figure below showing locating ring, center lines (H & V), leader pins and cavity locations. We can closely locate one mold half to other with the leader pins and side locks.

16 Cavity Mold w/ parting line locks

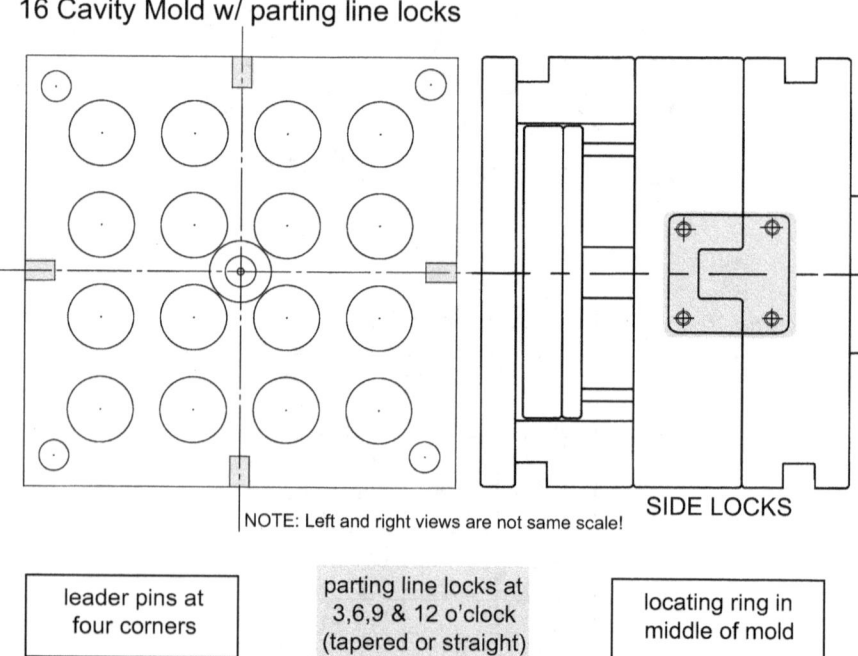

NOTE: Left and right views are not same scale!

SIDE LOCKS

leader pins at four corners	parting line locks at 3,6,9 & 12 o'clock (tapered or straight)	locating ring in middle of mold

NOTE: Side locks can be tapered or straight (use straight when there are vertical shut-offs or near vertical).

Alignment (in core/cavity stack)

If additional location is needed, then the various components in core, cavity, stripper (if required) stack can be interlocked via tapered fits incorporated into the various inserts (easier on round parts, but alignment wedges can often be incorporated into rectangular inserts as well. It is necessary for there to be some preload (0.0002 - 0.0005 inch) to accomplish positive location. Note: too much preload and mold is held open or insufficient clamp tonnage left for molding.....not enough preload and location is not optimized. The tapered fits should be tapered at an angle greater than 7° - 9° to avoid self locking or self holding tapers (e.g. Morse taper shanks for tool holders which are approx 5/8 inch per foot are less than 3° included angle); 12° - 15° angles are commonly used for good release in component stack alignment.

There must also be a slight clearance between core/cavity stack component and it's mold plate (insert OD to pocket ID) to permit insert to seek it's own optimum alignment. Obviously, the tapered alignment surface must be concentric with round molding surface in cavity or tapered wedge properly located with molding surfaces (in case of rectangular inserts).

In this example the cavity cannot float with the HR drop; thus, it will drive the mating components to align with it (cavity locates stripper ring which then locates core).

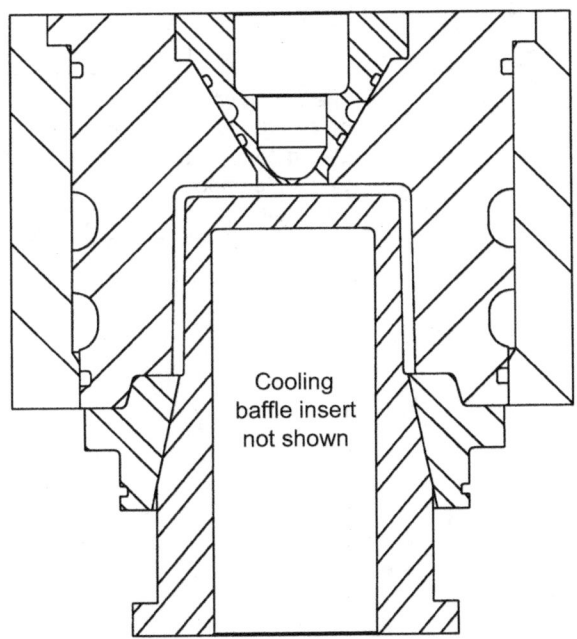

Cavity w/ interlocking stripper ring & core

If this stack was in the plate on previous page, a tapered sidelock would be sufficient .....actually with this construction, side locks would be redundant, but some cores and cavities will not permit such interlocking of stack components.

O-Rings: Circumferential Seals

When there are O-rings present in the cavity inserts, it is a good idea to step the pocket so that O-rings are not cut as they pass by intersecting coolant channels. In the case of the cavity stack seen previously, the cavity insert is stepped and pocket stepped to include a 30° lead in angle for inserting cavity insert with O-ring installed (remove all sharp corners).

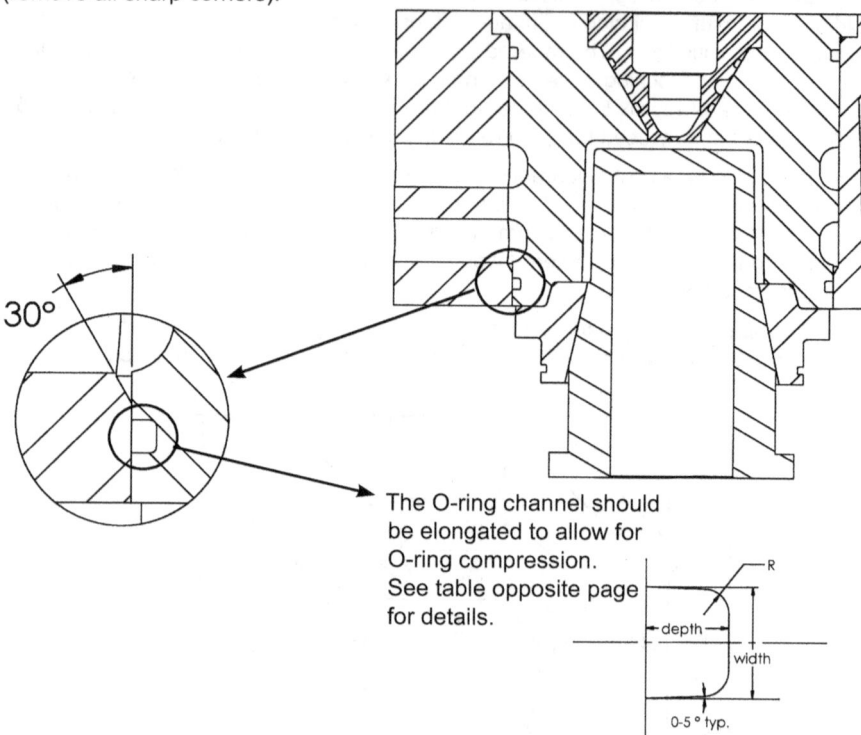

The O-ring channel should be elongated to allow for O-ring compression. See table opposite page for details.

O-Rings: Circumferential Seals

When there are O-rings present in the cavity inserts, it is a good idea to step the pocket so that O-rings are not cut as they pass by intersecting coolant channels. In the case of the cavity stack seen previously, the cavity insert is stepped and pocket stepped to include a 30° lead in angle for inserting cavity insert with O-ring installed (remove all sharp corners).

Design Guidelines: O-Rings in Circumferential Seals						
O-ring Number	Cross Size Inches	Actual Size Inches	Tolerance ± Inches	Groove Depth Inches	Groove Width Inches	Groove Radius Inches
2-004 to 050	1/16	0.070	0.003	0.050 to 0.052	0.093 to 0.098	0.005 to 0.015
2-102 to 178	3/32	0.103	0.003	0.081 to 0.083	0.140 to 0.145	0.005 to 0.015
2-201 to 284	1/8	0.139	0.004	0.111 to 0.113	0.187 to 0.192	0.010 to 0.025
2-309 to 395	3/16	0.210	0.005	0.170 to 0.173	0.281 to 0.286	0.020 to 0.035
2-425 to 475	1/4	0.275	0.006	0.226 to 0.229	0.375 to 0.380	0.020 to 0.035

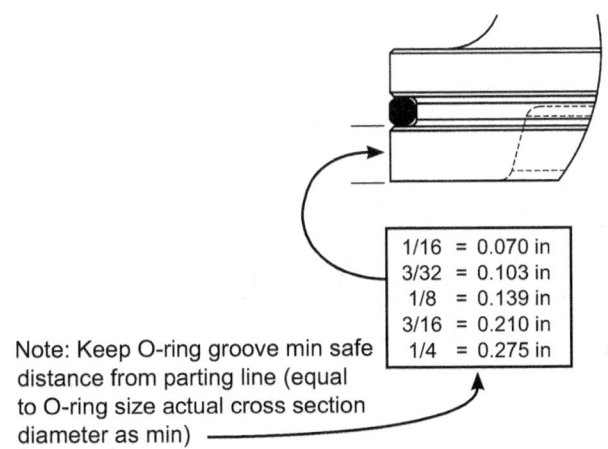

1/16 = 0.070 in
3/32 = 0.103 in
1/8 = 0.139 in
3/16 = 0.210 in
1/4 = 0.275 in

Note: Keep O-ring groove min safe distance from parting line (equal to O-ring size actual cross section diameter as min)

O-Rings: Face Seals

When sealing coolant channels which pass from plate to plate or plate to insert the preferred o-ring recess is per the figure below. This construction prevents the o-ring from slipping or extruding into the water channel. Where space does not permit use of the larger diameter required, then a tapered bottom can be used (see figures below). Note: The table below for face seals includes slightly different groove widths and depths versus groove for circumferential seals found on previous page.

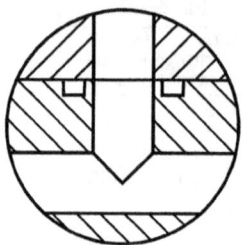

O-Ring Face Seal w/ Barrier Rib
(prevents O-ring from slipping into coolant channel)

O-Ring Face Seal w/ Angled Bottom
(helps prevent O-ring from slipping into coolant channel...when space does not permit design above)

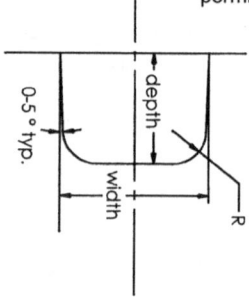

Design Guidelines: O-Rings in Face Seals						
O-ring Number	Cross Size Inches	Actual Size Inches	Tolerance ± Inches	Groove Depth Inches	Groove Width Inches	Groove Radius Inches
2-004 to 050	1/16	0.070	0.003	0.050 to 0.054	0.101 to 0.107	0.005 to 0.015
2-102 to 178	3/32	0.103	0.003	0.074 to 0.080	0.136 to 0.142	0.005 to 0.015
2-201 to 284	1/8	0.139	0.004	0.101 to 0.107	0.177 to 0.187	0.010 to 0.025
2-309 to 395	3/16	0.210	0.005	0.152 to 0.162	0.270 to 0.290	0.020 to 0.035
2-425 to 475	1/4	0.275	0.006	0.201 to 0.211	0.342 to 0.362	0.020 to 0.035

Cavity Pressure Transducer Installation Notes

Instructions for a standard 9211 transducer or equivalent. Design criteria should apply to most transducer installations – change dimensions as needed. The design guidelines help insure the transducer is not preloaded by square hitting a round, etc. as needed to provide good operating clearances and location.

Drawn: Jay W. Carender

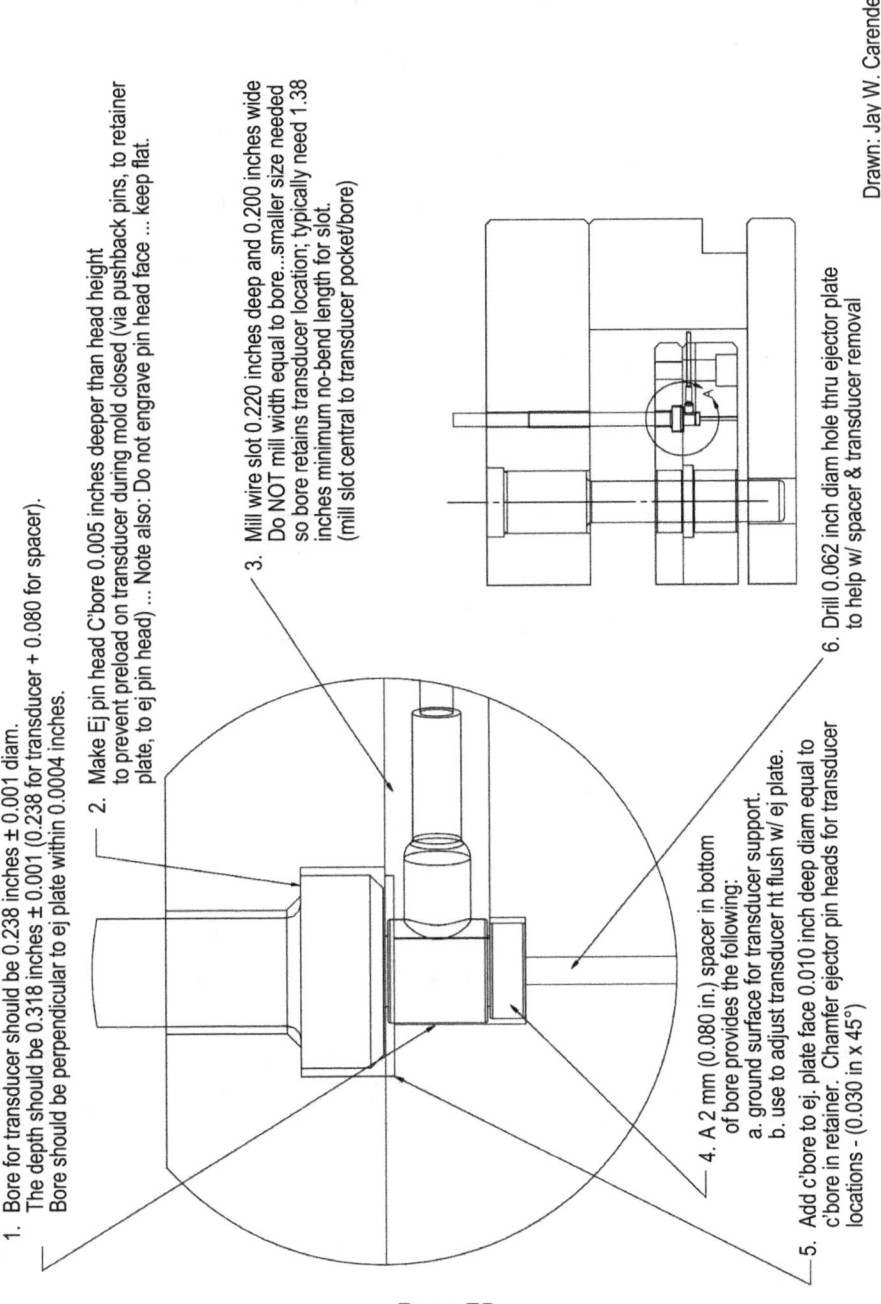

1. Bore for transducer should be 0.238 inches ± 0.001 diam.
   The depth should be 0.318 inches ± 0.001 (0.238 for transducer + 0.080 for spacer).
   Bore should be perpendicular to ej plate within 0.0004 inches.

2. Make Ej pin head C'bore 0.005 inches deeper than head height to prevent preload on transducer during mold closed (via pushback pins, to retainer plate, to ej pin head) ... Note also: Do not engrave pin head face ... keep flat.

3. Mill wire slot 0.220 inches deep and 0.200 inches wide
   Do NOT mill width equal to bore...smaller size needed so bore retains transducer location; typically need 1.38 inches minimum no-bend length for slot.
   (mill slot central to transducer pocket/bore)

4. A 2 mm (0.080 in.) spacer in bottom of bore provides the following:
   a. ground surface for transducer support.
   b. use to adjust transducer ht flush w/ ej plate.

5. Add c'bore to ej. plate face 0.010 inch deep diam equal to c'bore in retainer. Chamfer ejector pin heads for transducer locations - (0.030 in x 45°)

6. Drill 0.062 inch diam hole thru ejector plate to help w/ spacer & transducer removal

Torque Specifications For Fasteners

MATERIAL GRADE BOLT SIZE	SAE 2 MILD STEEL	SAE 5	SAE 8	SHCS	BRASS	SS AISI 303
GENERAL TORQUE SPECIFICATIONS ENGLISH FASTENERS (FOOT-POUNDS)						
1/4-20	6	11	12	13	5	5
1/4-28	7	13	15	16	6	7
5/16-18	13	21	25	27	8	9
5/16-24	14	23	30	33	9	10
3/8-16	23	38	50	52	15	17
3/8-24	26	40	60	60	16	18
7/16-14	37	55	85	86	23	25
7/16-20	41	60	95	95	25	28
1/2-13	57	85	125	130	32	37
1/2-20	64	95	140	145	34	40
9/16-12	80	125	175	180	44	50
9/16-18	91	140	195	210	48	54
5/8-11	111	175	245	255	68	75
5/8-18	128	210	270	290	73	80

GENERAL TORQUE SPECIFICATIONS
METRIC FASTENERS (NEWTON METERS)

MATERIAL CLASS MM-DIAM	4.6	4.8	5.8	8.8	9.8	10.9	12.9
5	3	4	5	7	8	11	12
6	5	6	8	12.5	14	17	20
6.3	5.5	8	9.5	14	16	21	24
8	12	16	20	30	34	44	50
10	23	32	40	60	70	85	100
12	40	56	70	103	120	150	180
14	65	90	110	167	190	240	280
16	100	140	170	270	290	380	440
18	137	177	225	350	--	480	580
20	200	--	330	520	--	740	860

Note: check also torque recommendations from your fastener supplier and/or equipment/product manufacturer for item which fasteners are being used. These values are approximate. SHCS used for Husky, Moldmaster and other hot runner systems may use slightly lower torque values; see supplier guidelines.

SHCS Dimensions (typical 1960 series)

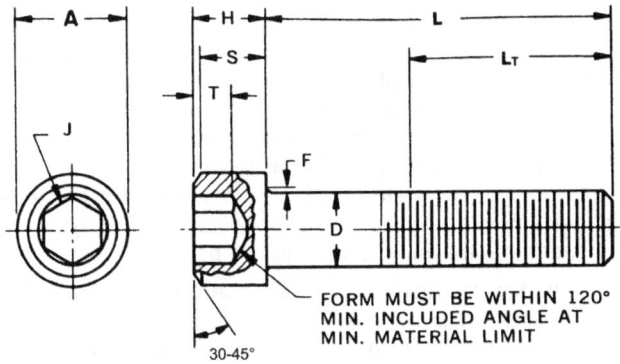

FORM MUST BE WITHIN 120°
MIN. INCLUDED ANGLE AT
MIN. MATERIAL LIMIT

30-45°

SIZE	D (MAX)	A (MAX)	H (MAX)	S (MIN)	J (NOM)	T (MIN)	F (MAX)	L (MIN)
0	0.060	0.096	0.060	0.054	0.050	0.025	0.007	0.50
1	0.073	0.118	0.073	0.066	1/16	0.031	0.007	0.62
2	0.086	0.140	0.086	0.077	5/64	0.038	0.008	0.62
3	0.099	0.161	0.099	0.089	5/64	0.044	0.008	0.62
4	0.112	0.183	0.112	0.101	3/32	0.051	0.009	0.75
5	0.125	0.205	0.125	0.112	3/32	0.057	0.010	0.75
6	0.138	0.226	0.138	0.124	7/64	0.064	0.010	0.75
8	0.164	0.270	0.164	0.148	9/64	0.077	0.012	0.88
10	0.190	0.312	0.190	0.171	5/32	0.090	0.014	0.88
1/4	0.250	0.375	0.250	0.225	3/16	0.120	0.014	1.00
5/16	0.312	0.469	0.312	0.281	1/4	0.151	0.017	1.12
3/8	0.375	0.562	0.375	0.337	5/16	0.182	0.020	1.25
7/16	0.437	0.656	0.438	0.394	3/8	0.213	0.023	1.38
1/2	0.500	0.750	0.500	0.450	3/8	0.245	0.026	1.50
5/8	0.625	0.938	0.625	0.562	1/2	0.307	0.032	1.75
3/4	0.750	1.125	0.750	0.675	5/8	0.370	0.039	2.00
7/8	0.875	1.312	0.875	0.787	3/4	0.432	0.044	2.25
1	1.000	1.500	1.000	0.900	3/4	0.495	0.050	2.50

Note:
1. 1960 series SHCS are typically made from a high grade alloy steel, hardened to a range of 37-45 RC.
2. "F" above is a fillet extension beyond "D".
3. Consult supplier to determine available lengths "L" for each screw size. Typical length increments are as follows:
   1/16" increments - lengths 1/8" thru 1/4"
   1/8" increments - lengths 1/4" thru 1"
   1/4" increments - lengths 1" thru 3.5"
   1/2" increments - lengths 3.5" thru 7"
   1" increments - lengths 7" thru 10"

Hot Runner Systems

HR Systems – There are many good system and controller suppliers available today. Certain systems may have specific advantages in various applications; thus, I will not endorse any specific system over another so as to not mislead readers. I do feel that some systems are much more reliable then others, but possibly at a <u>small</u> sacrifice in optimum temperature control, ΔP, etc. When the resin and/or part are simple or forgiving with regard to physical and/or dimensional requirements ... then use the most reliable, most simple and/or most cost effective system you can find. Some parts or resins demand sophisticated systems and controllers to deliver the best gate vestige and consistent melt delivery possible. Remember that your HR system is your melt delivery system: performance and repeatability are needed.

It is normally preferred for the manifold to be geometrically balanced. This means equal flow lengths of same sizes or combinations thereof to each cavity. Some layouts may require artificial balancing thru the use of CAD flow analysis to accomplish the desired manifold volumes, mold size, or other requirements for the specific application (artificially balanced means not equal length, but diameters changed to vary ΔP to each cavity in attempt to balance the fill[1]) ... these can work well in certain applications if properly designed.

Many mold builders today often purchase the entire HR system as a "bolt-on" system which includes not only the manifold and probes or nozzles, but also the manifold hsg, "A" half clamping plate w/ locating ring, and cavity support plate. Then the mold builder just bolts that purchase to their "A" half cavity retainer plate with cavities. Of course, proper coordination and sharing of mold layouts is required early in the design/build process.

It is generally best to have individual control (heater & T/C) for each cavity or probe drop in the system.

It is also useful to configure plates and mounting bolts so that the "A" half cavity plate and support plate can be transferred to the "B" half. This exposes the HR probe tips for cleaning and/or repair in the press. If and when such transfer takes place, the HR drops should be cooled to the mold temperature so that diametric seal surfaces are not in interference (otherwise seal surface damage may result and compromise seal). See figures on next pages.

---

[1] Remember we always want perfect balance of fill....when one cavity becomes full, they all should become full. This is the goal, but usually never fully achieved, but we do want to minimize any imbalance that does exist. We do not want cavities still only half full when first cavity is 98% full. Even equal flow length, geometric balanced runners do not achieve 100% balance. The dynamics of a hot gate opening the same every time are difficult to control with each and every shot; thus, it is always a good idea to run 25 consecutive short shots whereby only one cavity is nearly full and others are still short to see if the imbalance is repeatable – it may not be, and may not be changeable. If this is case, then do not use cavity pressure transducers for transferring the VP transfer on molding press. Also never attempt to balance by varying the gate diameter as shear rate will change resulting in increased variation. In artificially balanced systems the ID of channels prior to gate are adjusted as needed.

The aforementioned dynamics thru the hot gate are very different because the gate and runner are already filled with plastic. At the gate there is resin that must be kept hot, yet mold is trying to cool gate for best vestige and operation. Once one cavity opens, the pressure may drop as flow begins which results in less pressure for neighboring cavities to open gates. The dynamics of this scenario versus a cold and empty runner filling with molten plastic are much more complicated.

Husky® Hot Runner Systems

The figure below shows typical construction which permits cleaning of nozzle tips in the molding press. Note the guide pin/bushings to align "A" half cavity plate with HR "bolt-on" system. Note also that this requires interface bolts to be located in the main parting line to permit "A" half cavity plate movement over to "B" mold half. These bolts are removed while mold is open; "A" half plate is supported by guide pins and attached hoist for added safety. Not shown in figure below are leader pins from "B" mold half which will align and support the transferred "A" cavity plate to "B" mold half during nozzle tip cleaning. As restated on next page, the separation must occur at reduced nozzle tip temperatures to preserve the diametric seal surface in cavity insert as seen in figure below.

Guide Pin Bushing   Guide Pin

Figures courtesy of:
Husky Injection Molding Systems
530 Queen Street S.
P.O. Box 1000
Bolton, Ontario L7E 5S5
Canada
Phone: (905) 951-5000
Fax: (905) 857-4692

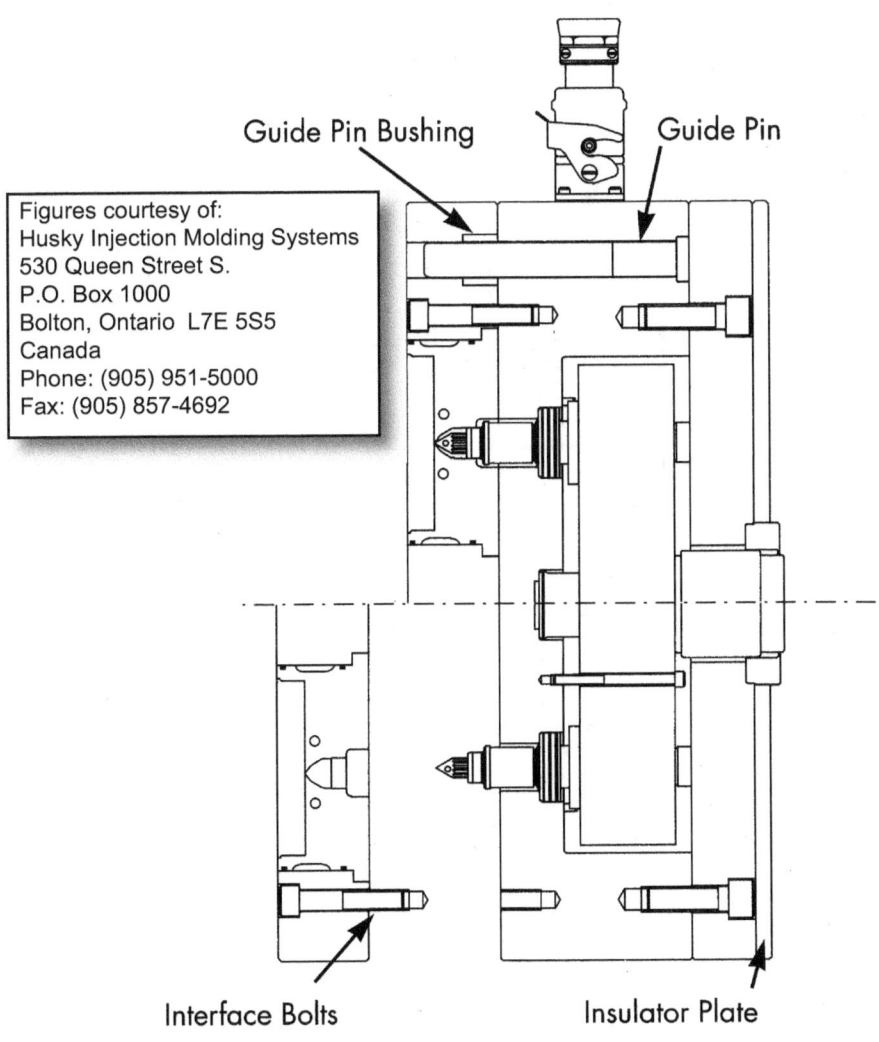

Interface Bolts   Insulator Plate

Husky® Hot Runner Systems (continued)

NOTE: The 0.7504 dimension in figure below will receive the nozzle. When the nozzle heats up to operating temperature, it will expand to make a tight seal so as to contain the high molding pressures and prevent leakage. Take care when splitting plates to clean nozzle tips in the press: the well in plate or cavity insert and probe temperature should be fully cooled down to room temperature to preserve the needed fit and finish during separation and reassembly (or at least be the <u>same</u> temperature in case of very hot molds). If performed at elevated nozzle temperatures AND colder well, the expanded nozzle housing OD may damage the hole ID leaving longitudinal lines where subsequent leakage might occur.

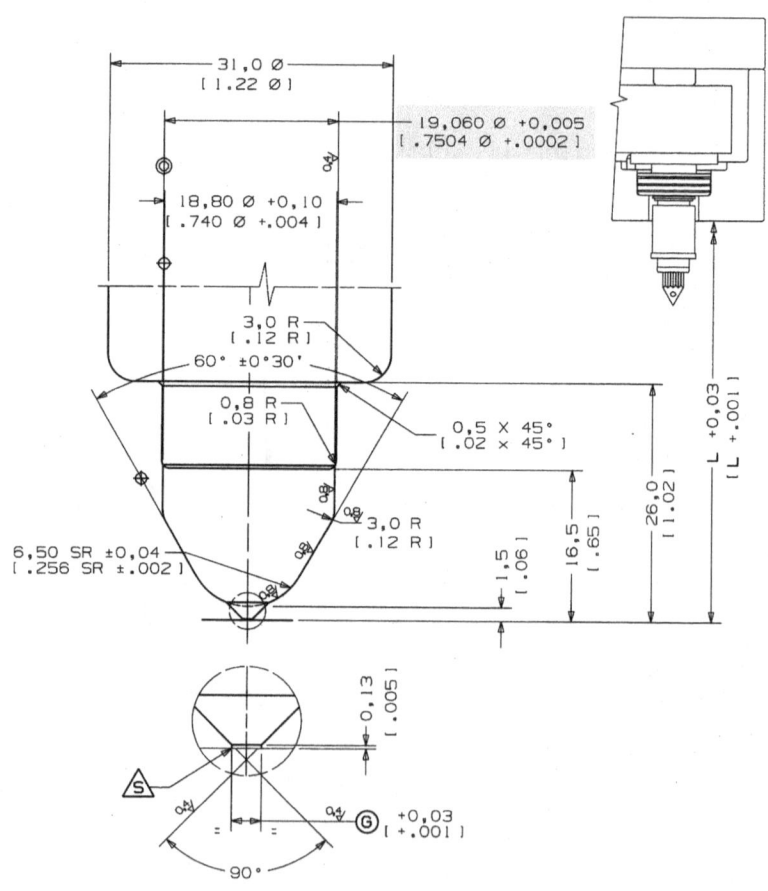

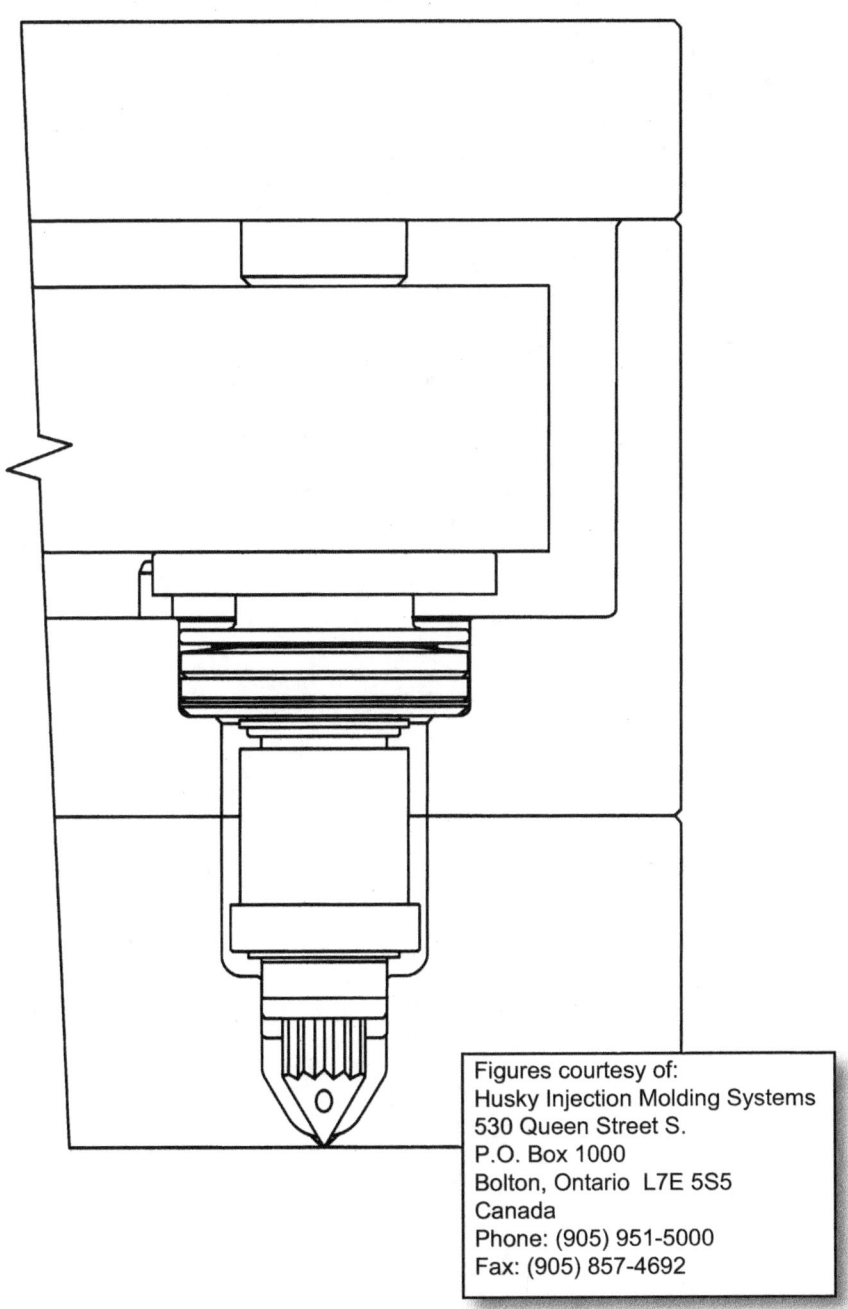

Figures courtesy of:
Husky Injection Molding Systems
530 Queen Street S.
P.O. Box 1000
Bolton, Ontario L7E 5S5
Canada
Phone: (905) 951-5000
Fax: (905) 857-4692

HR System Analysis & Advantages

You should always request an analysis sheet from your HR supplier. This shows the CAD calculated pressure drop, shear rate, residence time in HR system as well as showing the channel layout. This type documentation is very useful to project and process engineers.

The melt flow channels (hot runners) need to be as small as possible to minimize added residence time, but yet large enough to avoid excessive pressure drop. We also do not want to over shear the resin. Given these objectives, it is best to purchase from a reputable and skilled HR system supplier. The advantages of HR molds are as follows:

- Usually less total pressure drop thru runner
- Potential for reduced gate vestige without trimming
- Less or zero runner regrind
- Potential for faster cycle (offers less volume to inject, no runner to eject - less mold open time and cycle not limited by runner cooling)
- May be required for large cavitation molds (can use full hot or semi-hot)
- Full hot runner has no runner to separate - no separation losses
- Less projected area resulting in reduced clamp tonnage required
- Permits top center gating without use of three plate system (or other gate locations which may be difficult in cold runner)

MELT FLOW REQUEST SHEET

DATE: _____
REQUEST BY: _____

JOB INFORMATION:

MOLD NUMBER: _____
CUSTOMER: _____
PROJECT ENGINEER: _____
PART DESCRIPTION: _____
PART WEIGHT (g): __58__
DROPS/CAVITIES: __8 / 8__
RESIN GRADE: __SB 751__
RESIN TYPE: __COPOLYMER PP__
RESIN SUPPLIER: __Himont__
MELT TEMPERATURE (°C): __243__
MOLD TEMPERATURE (°C): __13__

JOB NUMBER: _____
SYSTEM DESCRIPTION: __750 HT__
WALL THICKNESS (mm): __.64__
TOTAL SHOT WEIGHT (g): __46.40__
HOT/COLD RUNNER: ☐ YES ☒ NO _____ (g/drc)
INJECTION TIME (s): __.25__ ☐ ACT ☐ ES
CYCLE TIME (s): __6.5__
MELT INDEX: __30__
MACHINE MODEL: _____
MACHINE INJECTION PRESS: _____
COLOR CHANGE: ☐ YES ☒ NO

COMMENTS: _____
_____
_____

DESIGN SPECIFICATIONS:

	DIAMETER (mm)	LENGTH(mm)/STD
SPRUE ORIFICE	8	
SPRUE BUSHING	11.5	80.
ANTI-DROOL		
M1	11.5	20
M2	11.5	201
M3	8	138.5
M4	8	20
M5		
M6		
NOZZLE HOUSING	8	90
NOZZLE TIP		535990
VALVE STEM		
GATE	.76 Ø	

MELT CHANNEL LAYOUT:

NOTE: - ASTERISK ✱ THE ABOVE DIMENSIONS OR COMPONENTS WHICH CANNOT BE CHANGED
- IF A GATE DIAMETER IS NOT SPECIFIED FOR HOT TIP JOBS PLEASE INDICATE HOW CRITICAL VESTIGE IS:

ANALYSIS RESULTS:

PRESSURE DROP (MPa): __21.0__
MINIMUM SHEAR RATE (1/s): __681__
TEMPERATURE RISE (°C): __5.0 c__
RESIDENCE TIME (s): __19.7s (3.a shots)__

SHEAR RATE @ GATE (1/s): _____
SHEAR STRESS @ GATE (MPa): _____
RESIN USED FOR ANALYSIS: __Himont__
__SD 242__

COMMENTS: _____

Hot Runner Connectors

Electrical Connectors - The EPIC® H-BE-24 connectors (24 pins/12 zones from Contact Electronics) is an excellent choice for use in hot runner systems to connect either the power or T/C wires. The EPIC® H-BE connectors have screw on type termination and can conduct up to 16 amps (pins are silver plated copper contacts). These connectors or equivalent are available from DME and other distribution sources; the point is to use the higher quality pins with screw on termination instead of crimped connectors.

Wiring should be as follows:

- Termination for power will be kept separate from T/C; Mold end termination for heater power should be male pins (results in power supply cable as female plug which is needed for safety).
- Mold end termination for T/C will be female sockets (needs to be opposite of power so always connected correctly). Pins 1 & 13 will be for the Sprue (or manifold zone #1 if sprue not present).
- All manifold zones will be next pairs of pins (e.g. 2 & 14, 3 & 15, etc until manifold is satisfied).
- Probes will start on the next available set of pins and continue until all probes are satisfied. Note: Probes will start on various pin/socket pair combinations from mold to mold - this is OK if so documented.
- T/C wires will be iron/constantan, iron is color coded white (+) and will be lower pins numbers (1-12). Not all countries use red and white for type "J" T/C, but iron will always attract to a magnet.
- The cables and cable plug/receptacle termination will typically be purchased from the controller supplier.
- The cable supplier must be aware of heater wattage per zone to provide proper wire sizes.
- "Coding pins" are available to safeguard against incorrect connections of identical plugs.
- Adopt some standard for numbering cavities and corresponding HR zone control such as: locating cavity one adjacent to leader pin offset corner; then, looking at the parting line, the numbers should count from top to bottom, then pick up with count from top to bottom of next column inward, etc until all cavs numbered.

NOTE: It is a fact that dissimilar materials act as a thermocouple and generate a small millivolt signal; it is generally thought that dissimilar materials in termination can be used because the error added on one side of the termination is subtracted by the other side of the termination in a given T/C wire type; thus, resulting in a correct final signal from the source T/C (i.e. In the termination, the iron to copper junction would be offset by the copper to iron mating junction). The aforementioned assumes that the temperature in the pin and socket are the same which is also generally accepted as true since the length is short and the material is copper which will enhance temperature equalization between the pin and socket. T/C wire should be used that is the same material as the thermocouple. Splicing in a long piece of copper wire would likely introduce error since different splice points would likely be different temperatures.

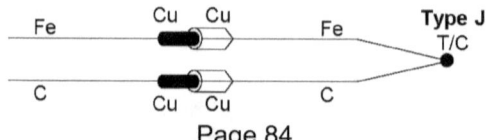

Page 84

Mold Design Checklist (P. 1 of 3)

A. Documentation

1. The final approved part drawing should be on file with mold builder/designer and project managers (includes part resin required).
2. Resin shrinkage rates used: flow and cross flow directions; shrink rate for each CTF dimension (critical to function) if available OR plan for steel safe planned recuts - nominalization plan and responsibility - the need for this depends on resin and part tolerances required.
3. Purchase order and mold specification list plus focal points for communication between molder and mold builder; ID project managers.
4. All components should be detailed sufficient to make replacement components without mold disassembly and measurement thereof.
5. The mold drawings should include a BOM (bill of materials) with altered items noted as such.
6. The end use molding press data should be on file with mold designer and builder including platen drawings listing platen size, tie bar clearance, min & max strokes, etc.
7. The mold design/drawings should be checked (by someone other than designer/detailer).
8. The cavity number layout and water circuit information should be listed on drawing (water circuit info to include such items as circuit ID number, channel size, component being fed (e.g. circuit #15 feeds cores 13-16 in series; 0.312" ID baffle drops).
9. Mold designer and/or project manager should document all phone conversations and ECRs and ECs; ECs should be approved by all.
10. There should be an approval process for the preliminary plan reviews and final mold drawings prior to mold construction.
11. Customers should insist on receiving a set of original mold drawings and/or a CAD database for the as-built mold.

B. Part Considerations

1. Is number of gates and their location OK, plus resulting probable weld lines?
2. Do part wall sections include thin to thick which may not pack properly?
3. Is there sufficient draft for release of wall depth and/or mold textures?
4. Will part geometry likely stick to ejector mold half as needed?
5. Will undercuts release without unacceptable deformation, skiving, etc?
6. Do lifters (or other actions) step in right direction to avoid skiving?
7. Will there likely be sink marks that may be objectionable?
8. Who is supplying the shrink rate (must be documented per above)? If the mold builder is responsible; is there sufficient knowledge relative to part tolerances (there are often various shrink rates for various dims)? Is there a need to build a unit tool to ID shrink rates and mold function?

C. Machine and Manufacturing Considerations

1. Will mold fit and function in the end use molding press:
   a. tie bar clearance: H & V (Horizontal & Vertical).
   b. mold height (hyd presses have a minimum; toggles have a min/max required.
   c. platen size and platen bolt patterns for mold clamps.
   d. min or max stroke problems; sufficient max daylight for open mold.
   e. ejector stroke and/or ejector force.
   f. KO pattern; need for KO extensions and/or pullback threaded holes.
   g. other: hyd core pulls; valve gates; pneumatic circuits; etc.
2. Part removal - Can parts & runner fall freely without hang-up or damage from misc mold projections? If robot unload: is there clearance for end of arm tooling; is there a need for ejection stroke positive stops; need for EOA locators on mold face such as tapered pilot hole recesses?
3. Is there ejection in part near gate so as to eject short shots?
4. If mold will run hot; is there a need for insulator plate on the clamp plates?
5. Does mold have correct sprue bushing radius & acceptable recess?
6. Will mold hang level with centrally balanced eyebolt hole?
7. Are air, water, electric & hyd connectors as needed for molding floor?
8. Are there SHCS counter bores or insert to pocket clearances in PL which will fill with plastic if PL flashes; can it be avoided?
9. Can gates be cleaned in the press (HR tools - transfer cavity plate?)
10. Is there a need for gate/runner shut-offs?

D. Mold Construction

1. Is the support plate thick enough and with enough support pillars?
2. Is locating ring recessed into clamping plate?
3. Do leader pins enter bushings before core enters cavity?
4. Are plate edges chamfered and do plates have sufficient eyebolt holes for plate handling (one middle of each side suggested on plates weighing over 25 lbs)?
5. Does mold include inserts for areas which will likely wear or break?
6. Will part geometry likely stick to ejector mold half (also B4 above)?
7. Do cores and cavities have detail #, cav #, hardness & mat'l engraved?
8. Will sliding steel surfaces gall? (4 pt min hardness diff; 8 is better; can use diff materials and/or surface treat components to achieve).
9. Are leader pin holes vented; will they catch parts (if so added clearance at bottom of hole is needed and/or provisions for parts to fall/clean out).
10. Is there an offset leader pin & return pin to insure correct mold assm? Are inserts oriented by shoulders, etc. to insure proper assembly?
11. Are there sufficient prybar slots?
12. Have parting line alignment blocks been specified or needed? (Are they correct type and location – straight or tapered and at center lines)?
13. Do passing shut-offs have max allowed angle & sufficient angle (7° minimum; 9° or more preferred)?
14. Do cavities or cores in blind pockets need jackscrews (cavities requiring removal from PL while in press)?
15. Are runner intersections radiused? Is there a coldwell?

Mold Design Checklist (P. 3 of 3)

16. Do slides have sufficient over travel? Are horn pins long enough?
17. Do mechanical slides need springs or other forms of retention to hold in place during mold open?
18. Is there a need for limit switches on slides or ejection return?
19. Is there adequate venting? Note: Can include perimeter venting; runner venting; inserts at critical internal features or internal last point of fill; self cleaning vents in ejector blades or pins, etc. Do vent relief channels vent to edge of mold (must vent to atmosphere)?
20. Are cavities and heel blocks sufficiently strong to withstand molding pressures? (not too close to edge; heel blocks must be backed up).
21. Are there ejector pins contacting the parting line? If so, are there provisions for early return such as spring loaded return pins?
22. Are ejector pins flush or below OR flush or above as needed?
23. Do stripper rings or ejector sleeves push on at least 75% of available wall surface area?
24. Can mold sit on floor without removing hardware; are feet required?
25. Are water lines at least ¼" away from ejector pins & 1xD from surfaces? (may need to be ½" away from pins on deeper holes - over 12" deep).
26. Is the ejector system guided? (guide pins optimally located).
27. Is the ejector plate thick enough relative to # of clearance holes and KO pattern? (small KO pattern on long plate may need thicker plate).
28. Are contoured ejector pins keyed as needed? (proper method).
29. Can ejector pins shift as needed in ejector plate to be self aligning and compensate for thermal expansion differences between ejector plate and core/cavity? (C'bore & hole clearances or coolant line in ejector plate).
30. Is there adequate cooling for runner system? (or manifold housing).
31. Will O-rings shear during component insertion/assembly?
32. Are O-ring grooves of optimal design to prevent mislocation?
33. Are O-rings in optimal location for ease of assembly? (plate vs insert).
34. Should steel with water lines be plated with electroless nickel?
35. Is there adequate cooling for the part: cores, cavs, inserts, pins, etc?
36. Is there a PL strap?
37. Do hydraulic cylinder activated slides need locks; are cylinders large enough?
38. Is there a need for spare stacks, T/C's, other components?

Mold Evaluation Options

Debug - This test is a basic check of mold functionality including the following:
- Set and run mold
- Establish a process and record for future reference (includes gate seal)
- Obtain sample parts; distribute as needed
- Perform basic checks (including wall thickness variability for similar walls)
- Suggest tooling improvements needed to improve processability (in writing)

Gate Seal[1] **-** Should be performed when first establishing the process during debug. Failure to perform this test may cause molding defects associated with under packing the parts such as sinks, excessive shrinkage and warpage. This test will identify the needed injection forward time; when this is estimated it may be estimated too low and cause the aforementioned problems and increase variability if gate does not become sealed.

Balance (Balance of Fill Analysis)[1] **-** A check of a multi-cavity mold's maximum imbalance of fill (part weight of least filled part versus first filled part; performed as a short shot). Processability is improved and variability is reduced with a more balanced fill.

DOE (Design of Experiments) - A DOE serves to check the mold's (and molded product's) sensitivity to process changes. A DOE done with a one cavity unit mold eliminates the "within group" variation coming from cavity to cavity variation; thus, improving the significance of process variation. With a one cavity mold the important process changes are not clouded by the aforementioned cavity to cavity variation. We want to identify what effect the process has on critical dimensions. The report then serves as a road map to guide processors toward what process variable to change to achieve necessary dimensional compliance.

Fill Time and Pressure Analysis (Relative Viscosity Testing)[1] **-** This test may be performed when there is concern for the availability of sufficient pressure to fill and pack part on the final production molding press. High pressure and large volumes of oil are needed when filling large cavitation molds. A debug press running a one cavity mold will easily display a performance edge versus a high cavitation mold. A spreadsheet can be arranged comparing the debug presses for unit and production molds versus the end use production press. On the spreadsheet the max fill rate and pressures for a given shot size can be compared. If a very fast fill rate or high filling pressure is required AND the spreadsheet indicates this is NOT doable in the production press, THEN the relative viscosity curve generated in this testing process will indicate where on the viscosity curve the production mold will operate. This information may be needed to plan for a different more suitable press selection OR identify the need to redesign the melt delivery system, gate location or plan a lesser number of cavities in the mold.

2-24 Hour PC Studies - This test would typically be done with the production mold in its final stage prior to normal production. If a required Cpk is planned, then it will be identified during this PC study run. Measurement time can become excessive; thus, a few critical dimensions should be identified for measurement. A common request would be an 8 hour run. This should be done using the planned production molding press.

---

[1] This test is outlined in the *IM Reference Guide, 4th Ed* & *IM Troubleshooting Guide, 3rd Ed.* The Gate Seal & Balance of Fill tests are also included in this book on pages 94 & 95.

Resin Process Temperatures

PROCESS MATERIAL	MELT TEMP (°F)	MOLD TEMP (°F)
ABS....med, high impact	440-510	90-180
ABS....high heat grade	510-540	140-180
ACETAL	380-420	140-220
ACRYLIC	420-485	120-180
CA, CAB	360-440	80-130
ETHYLENE VINYL ACETATE	350-400	50-100
IONOMER	420-460	50-100
NYLON....6/6	500-560	120-180
NYLON....6	470-530	100-180
NYLON....6/10	460-510	100-180
NYLON....11	420-480	80-140
NYLON....12	400-450	80-140
POLYCARB....lo/med viscosity	540-570	160-200
POLYCARB....high viscosity	590-640	180-240
POLYESTER - PBT	460-490	100-140
POLYESTER - PET (Bottle)	520-560	60-120
POLYESTER - PETG	480-520	70-120
POLYESTER - PCTG	520-560	70-120
POLYESTER - PCT (GF)	565-590	200-250
POLYESTER - PCTA (GF)	560-590	300-320
POLYETHERIMIDE	680-720	220-300
POLYETHYLENE....low density	340-440	50-100
POLYETHYLENE....med density	390-490	50-120
POLYETHYLENE....high density	420-540	50-150
PPO/STYRENE COPOLYMER	480-580	150-220
POLYPHENYLENE SULFIDE	590-670	190-230
POLYPROPYLENE	420-520	60-150
POLYSTYRENE....G.P.	380-460	50-120
POLYSTYRENE...impact modified	400-480	50-120
POLYSULFONE	650-750	200-320
POLYURETHANE	390-440	70-120
PVC....Flex - Rigid	320-420	50-120
SAN	400-500	120-180
T/P ELASTOMER	340-440	70-120

Notes:
1. Flame retardant grades can be 50° less.
2. Follow manufacturer's guidelines if available.
3. Glass or mineral fillers may require higher heats.
4. Amorphous polyesters may stick to hot steel > 150° F.
5. Long residence time may require lower barrel setpoints.

Setting and Starting The Mold

MOLD SETUP:

1. Determine resin requirements & availability; clean hopper, magnet, loader; load resin; start dryer if required.
2. Locate proper KO bars - all having equal length.
3. Prepare platens & mold using stone and mineral spirits.
4. Select eyebolt hole which yields a level hang/lift.
5. Do NOT stand below hanging mold; avoid hitting tie bars when lowering mold into place.
6. Line up locating rings; slowly close mold.
7. Level mold if not already and clamp to fixed platen.
8. Open moving platen (w/ hoist still attached/supporting mold); install KO bars. If KOs are acting as pullbacks: tighten bars making certain they bottom out against the ejector plate in mold.
9. Close platen; clamp mold to moving platen; remove safety straps; unhook hoist.
10. Open mold to desired daylight; set slowdown switches with certainty that banging the mold will not occur; fine tune the final switch positions by repetitive small adjustments, observations and readjustment.
11. Secure KOs to ejector plate of press; set stroke.
12. Connect all required power - hydraulic, electric, pneumatic.
13. Make sure powered functions are functional; run electrical heaters just long enough to prove functionality avoiding excessive heat buildup before water is connected.
14. Connect water lines using an acceptable number of loops/jumpers; locate lines clear of any interference. Avoid having all "INS" on the same side.
15. Recheck fittings for proper connection; turn water on; (heaters should be off); look for leaks.

PROCESS SETUP (IF UNKNOWN):

16. Set barrel profile per resin supplier's recommended mid-range (same logic for mold temp).
17. Estimate the shot size and set machine for approximately 2/3 of the mold's full shot. Set decompression stroke. Set a position transfer point (if machine is so equipped) approximately one inch from bottom.
18. Estimate & set second stage time; set second stage pressure at zero.
19. Set 1st stage pressure at 50% for starters (this may ultimately be set at 100% - assuming Decoupled Molding[SM]).
20. Set velocity to maximum.
21. Estimate and set cooling time.
22. Set back pressure at 50 psi.
23. Refer also to Hot Runner Start-up procedure if applicable.

MOLD START-UP:

24. Purge barrel free of degraded resin.
25. Set machine for semi-auto; start cycle; observe screw.
26. Adjust velocity and/or pressure as needed; if the fill was fast and short as estimated, the pressure can be increased. The fill pressure should be set high enough so the fill speed is not pressure limited, but controlled by velocity setpoints. If flash or dieseling occur, slow the velocity.
27. After observing each cycle, the shot size and transfer point will be adjusted frequently to set the process so that the first stage accomplishes 95 - 98 % of the fill as measured by shot weight.
28. Once the first stage shot size, transfer, velocity and pressure are set, we can set 2nd stage packing pressure.
29. Adjust pack pressure as needed, but do not overpack.
30. Recheck cushion; some cushion should be maintained.
31. Set screw rpm so recovery is completed just prior to next cycle, but not limiting cycle time.

PROCESS DOCUMENTATION:

32. Record all basic machines setpoints on the setup sheet.
33. Note the transfer time (fill time) and weight.
34. Note the overall cycle time.
35. Note the ejection: multiple, push only, push/pull, etc.
36. Total shot weight, part weight, % runner, etc.

IMPORTANT NOTE: When using Decoupled Molding[SM] (see also pages discussing this technique): Considerable skill and specific mold and machine knowledge is required when setting pressures near maximum. Set pressures in accordance with consideration for mold damage in the event some parts do not shoot due to gate blockage and remaining cavities actually see the elevated 1st stage pressure.

Hot Runner Startup & Operation

Hot runner systems offer the following benefits to the molder: elimination or reduction of regrind; automation of part handling is easier; potential for faster cycles; part/runner separation is avoided; injection pressures may be lower. It is important to follow certain guidelines to get the most out of your hot runner system.

Some mold makers will list min and max temperature differentials for the manifold system versus the mold temperature; these should be followed as they allow for proper preload from thermal expansion. This is important to prevent leakage and/or excessive compression damage. A heat soak time of approximately thirty minutes should be allowed after set points are reached to permit proper thermal expansion (assumes resin w/ good thermal stability ... some will require less heat soak time). Front clamping plate cooling is necessary to prevent excessive heating of the platens. It is worthwhile to start-up the system with all zones off; then check each zone by itself to determine that the T/C response is for the respective zone heating. Always make a cavity sheet which indicates which zone number is what cavity and what section of mold relates to which manifold zone. Set the manifold temperature equal to the process melt temperature. Adjust drops as needed to achieve proper gate vestige. Chilled molds need to be started at room temperature. Do not connect gate coolant water in series. If your controller doesn't have soft start or ground fault capabilities; set the temps at only one hundred degrees for the first half hour so as to dry the absorbed moisture out of heater insulation. If a filter nozzle is used, great care is needed in filter nozzle selection to avoid excessive pressure drops - both on injection and ability to decompress manifold before mold opens. Always perform a balance of fill analysis before the mold leaves the mold maker; insist on a maximum imbalance of ten to twenty percent (depending on cavitation). It is best to never purge through the open hot runner mold due to potential for leakage and potential damage to cavities (i.e. some systems which have multiple cavities in a single cavity block could develop forces greater than retention screws retaining cavities or plate strength). It is necessary to sometimes purge through an open mold to purge degraded resin: try to purge with back pressure or minimal injection pressures; never use operating pressures to purge through the open mold. Never connect/disconnect plugs which are carrying power as it can damage most controllers besides being unsafe.

New Mold Debug Checklist

EASE OF MOLD LIFTING AND SETTING:
1. Will mold fit the end use molding press?
2. Mold should have eyebolt holes & safety strap.
3. Mold should hang level to align with locating ring & KOs.
4. Mold should offer protection to external wiring, switches, fittings, etc on bottom of mold.
5. Mold should have clamping slots positioned such that sufficient clamps can be used to secure mold to platens.
6. All water fittings should be positioned so that not in interference with mold clamps, machine doors, tie bars, etc.
7. Mold should have sufficient eyebolts so that mold rotation can be achieved with a second eyebolt and hoist.
8. Sprue radius should be compatible, sprue recess ID should be compatible with machine's nozzle/heater band OD.
9. Mold should have total weight listed on clamping plate.
10. Socket head cap screws retaining clamping plate should be slightly below flush to prevent platen damage.
11. Electrical receptacles should be accessible & away from water fittings.

BASIC MECHANICAL FUNCTIONALITY:
12. Water lines should be free flowing; check with flow meter/indicator or direct return into bucket to observe flow (do not try to measure/quantify with bucket – not accurate due to lack of back pressure).
13. Ejector plate should return freely and completely.
14. Slides should move freely, but be retained during mold open in the pulled position.
15. Leader pins should not exhibit galling.
16. All moving slide surfaces and slide locks should be free from galling.

ELECTRICAL FUNCTIONALITY:
17. Know the heater wattage; calculate the full load amperage and compare to actual (or at least compare to other like heaters).
18. Each T/C should indicate a proper response for it's heater.
19. Heater cables should be grounded at mold & controller ends.

LOCATION ITEMS:
20. Ejector pins should be flush to 0.001" below flush.
21. Examine parts looking for long ejector pins.
22. Examine parts/runner looking for parting line mismatch.
23. Examine parts looking for slide mislocation.
24. Check part for proper wall thickness with pointed micrometers.
25. Examine wall thickness variation relative to fill problems.
26. Intentional drags should be adjacent to ejector pins and be shallow @ 0.003", but sharp (can then reduce sharpness if too much drag).
27. Alignment should be located on horizontal and vertical center lines.

ESTABLISH A BASIC PROCESS:
28. Record all pertinent process & setup data.
29. Determine gate seal time.

DETERMINE BALANCE OF FILL:
30. Establish basic process.
31. Set hold/pack pressure to zero.
32. Adjust inject time or transfer position to achieve only one full part (or as close as possible).
33. Collect 3 shots; separate parts by cavity and determine avg part weight for each cavity.
34. Max imbalance should be 15% or less.

MISC OTHER CHECKS:
35. Have all sharp corners been removed from mold base exterior?
36. Do parts fall free and clear of mold without nicks?
37. Have pry bar slots been installed in main parting lines?
38. Review gate vestige relative to requirements.
39. Review for proper mold surface finish from both cosmetic and functionality standpoint.
40. Lightly touch each core and cavity just after repeated cycling looking for hot spots (Do not touch hot molds; observe standard safety practices before reaching into press).
41. Stop mold prior to any ejection, look for raised surfaces where part may be trying to stick in non-moving half.
42. Look at parts for parting line drags.
43. Perform dimensional checks after shrinkage (48 hrs).

Root Cause Analysis

Control limit violations should result in some form of root cause analysis and documentation. When performing such root cause analysis, it is worthwhile to understand all the inputs to the process – the many variables affecting the process. Below is a listing of the many items to check, and they are listed in the approximate order of importance and/or ease of checking.

Process Checks for Root Cause Analysis
1. Verify that the part has been remeasured; rule out measurement.
2. Proper shot size and resulting cushion.
3. Mold temperature control units are on and at proper setpoints.
4. Pack/hold pressure is at proper setpoint.
5. Barrel temperature setpoints are at proper setpoint.
6. Back pressure is turned on and at proper setpoint.
7. VP transfer (boost cutoff) is at proper setpoint (and mode thereof).
8. If resin is hygroscopic (meaning – does it require drying):
    a. dryer is at proper setpoint temperature.
    b. dryer has not run low in recent hours affecting residence time.
    c. dewpoint is properly low (desiccant is good, bed regeneration is good, etc).
    d. dryer and/or material lines are not dirty with contamination.
    e. have maintenance check filters, regeneration heaters, desiccant, etc.
9. Injection forward timers are at proper setpoint.
10. Injection fill velocity is at proper setpoint resulting in proper fill time.
11. Cool/cure timer is at proper setpoint resulting in proper cycle time.
12. Hot runner setpoints are at proper setpoint.
13. Mold is watered correctly.
14. Clamp tonnage is sufficient for the mold.
15. Resin lot changes having stiffer or easier flow; different thermal properties affecting cooling, warpage, void formation, etc.
16. If the above checks good, have maintenance check the press operation: heater bands, hyd. valves, screw & check rings, etc.

Tooling Checks for Root Cause Analysis
NOTE: Many of the following can also be caused by process issues.
A. If part sticking or deforming:
    Check for parting line drags and rolled steel creating drags.
B. If part is flashing:
    Check for parting line damage, plastic beneath slides, blocked gates.
C. If part has FM or metal contamination:
    Check for galling ejector pins, other moving steel, vertical shut-offs, etc.
D. If gate vestige is poor such as stringing or high gates:
    Check for optimum gate size, check for smashed T/C wires, proper heater size and placement, gate lands, probe tip height for HR tools ... NOTE: depending on gate type, there may be many other possible issues causing poor gate appearances.
E. If part exhibits splay or water spots:
    Check for water leaks caused by damaged o-rings, cracked cores or cavities; check also for hose or pipe nipple leaks whereby the water can migrate thru pipe recesses to steel inserts, then onto molding surface – especially possible with water lines on top of mold.
F. If part exhibits flow lines:
    Check for plastic debris in gate drop from previous shot; check also for sharp corners, engraving, part/tool features affecting plastic flow.
G. If part exhibits pin push:
    Check for excessive puller depths, insufficient ejection or poorly placed ejection; check also for air poppets that are not functional.

Determining Gate Seal Time

Knowledge of the gate seal time, permits the molder to use the needed injection forward time to accomplish effective packing. Injection forward times less than the gate seal time often result in sinks, voids and increased shrinkage resulting in poor dimensional compliance. The total injection forward time should be set a fraction longer than the gate seal time if cycle time permits. Hot tip gates may not result in a seal due to localized packing around the hot gate. This packing is of little or no benefit; thus, do not attempt a full gate seal on some hot gates parts. Note also: valve gates are sealed mechanically... this test applies differently to valve gated molds; whereas, it can still indicate optimum injection time, but valve must be closed in time to achieve good gate vestige.

Make a plot of total part weight vs increasing increments of pack and/or hold time after VPT as shown in table at right.

Increments can be ½ or 1 sec ... or finer increments as needed. Bigger parts with bigger gates need bigger increments of time.

When the part weight becomes stable, the gate is sealed and no more pack/hold time is needed.

Unlike the Balance of Fill test on next page ... here the graph (below) is very helpful to understand the data!

Hold Time (sec)	Shot Weight (parts only) (grams)
0.60	8.716
0.70	8.770
0.80	8.810
0.90	8.830
1.00	8.850
1.10	8.863
1.20	8.878
1.30	8.885
1.50	8.902
1.70	8.913
1.80	8.915
2.00	8.916
2.20	8.916
2.40	8.916

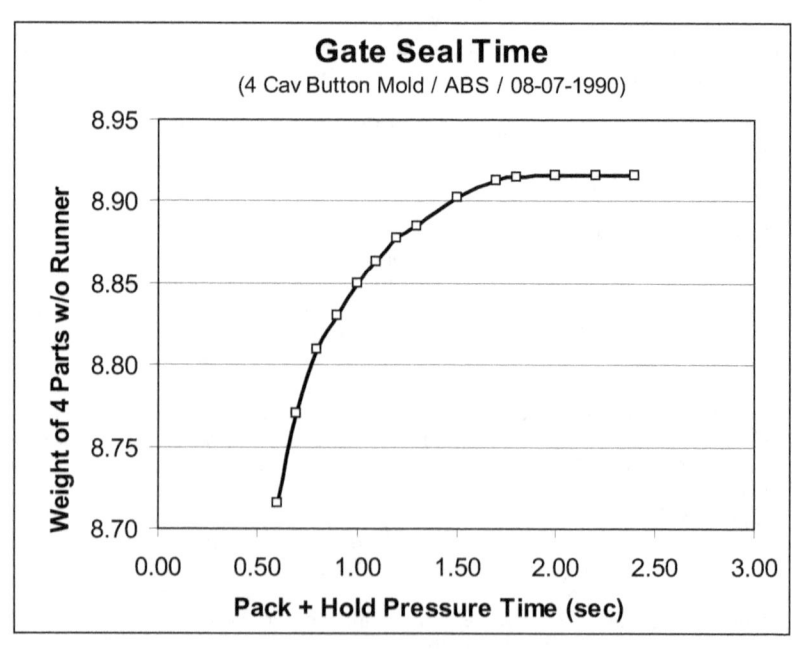

Gate Seal Time
(4 Cav Button Mold / ABS / 08-07-1990)

Balance of Fill Analysis

Cavity	Avg Part Wt (grams)	Fill Seq	Avg % Unbal	% Full	Part wts (grams) shot 1	shot 2	shot 3
4	1.40	1	0.00%	100.00%	1.42	1.39	1.38
5	1.39	2	0.72%	99.28%	1.39	1.39	1.38
12	1.38	3	1.19%	98.81%	1.39	1.37	1.38
13	1.38	4	1.43%	98.57%	1.37	1.39	1.37
6	1.37	5	1.91%	98.09%	1.38	1.37	1.36
14	1.34	6	3.82%	96.18%	1.34	1.35	1.34
11	1.33	7	4.77%	95.23%	1.36	1.32	1.31
3	1.33	8	5.01%	94.99%	1.35	1.32	1.31
15	1.19	9	14.56%	85.44%	1.20	1.21	1.17
10	1.15	10	17.66%	82.34%	1.16	1.15	1.14
9	1.14	11	18.38%	81.62%	1.15	1.15	1.12
7	1.12	12	19.57%	80.43%	1.12	1.14	1.11
1	1.11	13	20.29%	79.71%	1.13	1.11	1.10
8	1.06	14	23.87%	76.13%	1.10	1.05	1.04
16	1.06	15	23.87%	76.13%	1.07	1.06	1.06
2	1.03	16	26.01%	73.99%	1.04	1.03	1.03

PROCEDURE:

1. Process should be running at equilibrium with regards to mold & melt temperature, pressures normal & stable.
2. Consider mold's ability to run a short shot.
3. Set feed or transfer to run short shots whereby one part is full (or nearly so); begin shot collection. Typically, this will be 1st stage fill only as we are only concerned with balance of FILL ... we do not need any pack or hold psi in this test.

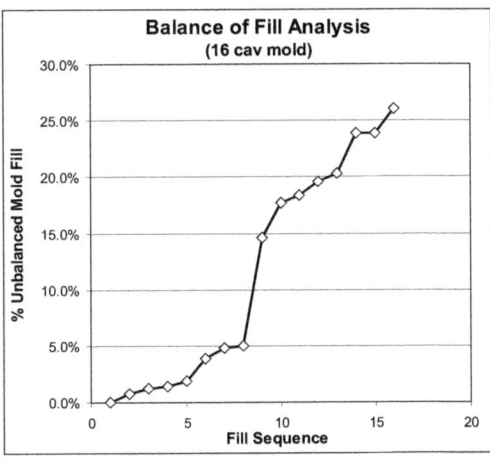

4. Collect three shots separated by cavity number, weigh individual parts and record, calculate average.
5. Arrange data in table, sorted in descending order (from high to low) of average part weight.
6. Compute % unbalanced (this mold has a 26% imbalance - see table above).
7. Restore process to normal.
8. Graph serves no purpose other than display of data.

Sample Mold Information Sheet

Mold Information Sheet

MOLD DESCRIPTION					
NUMBER OF CAVS.		MOLD DRWG NUMBER			
PART DESCRIPTION					
RUNNER TYPE					
MOLD SIZE	HORIZ	VERT	SHUT HT	WT	

SPRUE RADIUS		SPRUE ORIFICE		
RECESS I.D.		RECESS DEPTH		
EJECTION TYPE				
K.O. PATTERNS				
K.O. THREAD				
GATE SIZE/TYPE				
CAVITY MATERIAL		HARDNESS		Rc
CORE MATERIAL		HARDNESS		Rc
POWER	MANIFOLD			
CONSUMPTION (watts)	GATE DROPS			

SEE CAVITY LOCATIONS BELOW
TOP/EJECTOR HALF   00

```
OPERATOR SIDE

   13      9        5      1

   14     10        6      2

      Manifold 2        Manifold 1

   15     11        7      3

   16     12        8      4
```

Adopt a standard cavity numbering sequence such as top to bottom starting at offset corner (back side, top in this example).

Since the HR system is located in A-half, the layouts typically start in top left when looking at parting line; thus, this layout. This view is B half where you grab the parts!

Sample Mold Process Sheet

Molding Conditions Record

DATE/TIME: _____   PART: _____   PART WT (GRAMS): _____

PART DESCRIPTION: _____   MOLD NO: _____   SHOT WT (GRAMS): _____

REV # OR ECN (IF APPLICABLE): _____   NO. CAVITIES: _____

MACHINE ID: _____   TONS: _____   SHOT SIZE (OZ): _____

MECH ADV: _____   SCREW TYPE: _____   NOZZLE ORIFICE ID: _____

RUN BY: _____

CORE SEQUENCE: _____   CIRCLE EJECT TYPE: PUSH ONLY   PUSH/PULL   PUSH/SPRING RETURN   OTHER

	RUN NO. OR ACTIVITY DESCRIPTION >>>>>							
MATERIAL	MATERIAL							
	LOT NO.							
	COLOR CONC							
	COLOR BLEND RATIO							
	OTHER ADDITIVES							
DRYING	DEW POINT	(°F) (°C)						
	TEMPERATURE	(°F) (°C)						
	TIME	(HOURS)						
	HOPPER SIZE							
	CALCULATED NORMAL RESIDENCE TIME							
TEMPERATURES	FEED ZONE	(°F) (°C)						
	CENTER ZONE	(°F) (°C)						
	FRONT ZONE	(°F) (°C)						
	NOZZLE	(°F) (°C)						
	HOT MANIFOLD	(°F) (°C)						
	ACTUAL MELT:	(°F) (°C)						
	MOLD - FIXED	(°F) (°C)						
	MOLD - MOVEABLE	(°F) (°C)						
	MOLD - SLIDES	(°F) (°C)						
PRESSURE	FILL VELOCITY	(%)						
	CLAMP	(%) (TONS)						
	FILLING PRESSURE	P1						
	PACKING PRESSURE	P2						
	HOLD PRESSURE	P3						
	BACK PRESSURE							
CYCLE TIMES	INJ, TOTAL SCREW FORWARD	(SEC)						
	FILL TIME	(SEC)						
	PACK TIME	(SEC)						
	HOLD TIME	(SEC)						
	COOLING	(SEC)						
	PLASTICIZING	(SEC)						
	MOLD OPEN TIMER	(SEC)						
	ACTUAL TOTAL OPEN	(SEC)						
	OVERALL	(SEC)						
	RESIDENCE TIME - BARREL	(MIN.) CALC						
MISCELLANEOUS	TRANSFER METHOD							
	TRANSFER POSITION	(INCHES) (MM)						
	TRANSFER WEIGHT	(%) CALC						
	DECOMPRESSION	(INCHES) (MM)						
	SHOT SIZE	(INCHES) (MM)						
	CUSHION	(INCHES) (MM)						
	END OF STROKE	(INCHES) (MM)						
	SCREW SPEED	(RPM) (%)						
COMMENTS	A -							
	B -							
	C -							
	D -							
	E -							

Shot to Shot vs Within the Shot Variation

Dimension = 1.037 = (Nominal)
Min Tolerance = 1.032
Max Tolerance = 1.042

COLLECT SHOTS EACH 5 MINUTES FOR 2 HOURS; PLACE IN LABELED BAG
RECORD PROCESS AND ANY CHANGES MADE THROUGHOUT THE RUN (THERE SHOULD BE NONE)

CAV	1	2	3	4	5	6	7	8	9	10-22 (not shown)	23	24	AVG	STD DEV σ_{n-1}	Ppk	Pp
1	1.037	1.037	1.036	1.037	1.037	1.036	1.037	1.037	1.036	....	1.037	1.037	1.037	0.00048	3.23	3.46
2	1.038	1.037	1.037	1.037	1.037	1.037	1.038	1.037	1.037	....	1.038	1.037	1.037	0.00041	3.85	4.02
3	1.037	1.037	1.038	1.037	1.037	1.038	1.037	1.037	1.038	....	1.037	1.037	1.037	0.00048	3.23	3.46
4	1.037	1.037	1.038	1.038	1.037	1.038	1.037	1.037	1.038	....	1.037	1.037	1.037	0.00051	2.97	3.27
5	1.039	1.040	1.039	1.038	1.040	1.039	1.039	1.040	1.039	....	1.039	1.040	1.039	0.00066	1.41	2.53
				........cavities 6 thru 30 not shown........												
31	1.039	1.040	1.040	1.040	1.040	1.040	1.039	1.040	1.039	....	1.039	1.040	1.040	0.00044	1.70	3.77
32	1.039	1.039	1.039	1.038	1.039	1.038	1.039	1.039	1.039	....	1.039	1.039	1.039	0.00048	2.31	3.46
AVG	1.038	1.038	1.038	1.038	1.038	1.038	1.038	1.038	1.038	....	1.038	1.038	AVGs 1.038			
σ_{n-1}	0.0009	0.0012	0.0013	0.0011	0.0012	0.0013	0.0009	0.0012	0.0012	....	0.0010	0.0012	0.00115			
MIN	1.036	1.035	1.035	1.036	1.035	1.035	1.036	1.035	1.035	....	1.036	1.035	1.035			
MAX	1.039	1.040	1.040	1.040	1.040	1.040	1.039	1.040	1.040	....	1.040	1.040	1.040			
n	32	32	32	32	32	32	32	32	32	....	32	32	32			
Ppk	1.55	1.16	1.05	1.21	1.16	1.069	1.55	1.16	1.11	....	1.44	1.14	1.22			
Pp	1.77	1.39	1.31	1.56	1.39	1.33	1.77	1.39	1.37	....	1.66	1.36	1.47			

As can be seen from these averages, the variation is greater within the group or within the shot (or in other words: cavity to cavity variation). The shot to shot variation is less. There is potentially more opportunity to improve Cpk by working on the cavity to cavity variation. Additional study is needed to see if it is steel variation or caused by something else such as imbalanced fill or cooling differentials.

Average of shot to shot variation (between group variation; indicates process consistency) = 0.00060 σ_{n-1}
Average of cav to cav variation (within subgroup variation; may indicate cav steel variation, imbalance fill or temp differentials, etc) = 0.00115 σ_{n-1}
Average Ppk of 24 shots; target is 1.33 or greater = 1.22 Ppk

Calculating and Correcting Ppk

0.400 NOMINAL DIMENSION
0.403 MAX SPEC LIMIT
0.397 MIN SPEC LIMIT

CAV	SHOT 1	SHOT 2	SHOT3	AVG	MIN	MAX	RANGE
1	0.401	0.400	0.400	0.400	0.400	0.401	0.001
2	0.402	0.401	0.402	0.402	0.401	0.402	0.001
3	0.400	0.400	0.401	0.400	0.400	0.401	0.001
4	0.400	0.401	0.400	0.400	0.400	0.401	0.001
5	0.402	0.401	0.401	0.401	0.401	0.402	0.001
6	0.401	0.401	0.401	0.401	0.401	0.401	0.000
7	0.400	0.399	0.400	0.400	0.399	0.400	0.001
8	0.400	0.401	0.401	0.401	0.400	0.401	0.001
9	0.401	0.401	0.400	0.401	0.400	0.401	0.001
10	0.401	0.400	0.401	0.401	0.400	0.401	0.001
11	0.400	0.400	0.400	0.400	0.400	0.400	0.000
12	0.400	0.400	0.400	0.400	0.400	0.400	0.000
13	0.400	0.401	0.400	0.400	0.400	0.401	0.001
14	0.401	0.401	0.401	0.401	0.401	0.401	0.000
15	0.400	0.401	0.401	0.401	0.400	0.401	0.001
16	0.402	0.402	0.401	0.402	0.401	0.402	0.001
17	0.402	0.401	0.401	0.401	0.401	0.402	0.001
18	0.400	0.401	0.401	0.401	0.400	0.401	0.001
19	0.400	0.400	0.401	0.400	0.400	0.401	0.001
20	0.401	0.402	0.402	0.402	0.401	0.402	0.001
21	0.401	0.401	0.402	0.401	0.401	0.402	0.001
22	0.400	0.401	0.401	0.401	0.400	0.401	0.001
23	0.401	0.400	0.401	0.401	0.400	0.401	0.001
24	0.400	0.400	0.400	0.400	0.400	0.400	0.000
25	0.401	0.401	0.402	0.401	0.401	0.402	0.001
26	0.400	0.400	0.401	0.400	0.400	0.401	0.001
27	0.402	0.402	0.401	0.402	0.401	0.402	0.001
28	0.401	0.401	0.402	0.401	0.401	0.402	0.001
29	0.400	0.400	0.401	0.400	0.400	0.401	0.001
30	0.400	0.400	0.400	0.400	0.400	0.400	0.000
31	0.401	0.400	0.400	0.400	0.400	0.401	0.001
32	0.400	0.401	0.400	0.400	0.400	0.401	0.001
AVG	0.4007	0.4007	0.4008	0.4007			0.001
COUNT	32	32	32	32			32
MIN	0.400	0.399	0.400	0.399667			0.000
MAX	0.402	0.402	0.402	0.402			0.001
RANGE	0.002	0.003	0.002	0.002333			0.001
σ_{n-1}	0.000745	0.000701	0.000693	0.000713			0.000397
Pp	1.34	1.43	1.44	1.40			
		1.61	1.76	1.55	1.64 after adjustment - Note #4 below		
Ppk	1.05	1 12	1.05	1.072			
		1.32	1.45	1.20	1.32 after adjustment - Note #4 below		

NOTES:

1. Having uniform core & cavity sizing is beneficial from a mold maintenance standpoint; thus, review actual sizing before a change, then verify need for change by review of process and factors contributing to cavity to cavity variation.

2. Comparing three shot cavity averages with the nominal specification will indicate location for core or cavity in need of resizing.

3. A Pp significantly greater than the Ppk also indicates potential to improve Ppk.

4. The Ppk above could be improved to average nearly 1.33 by adjusting steel for cavities (2, 16, 20 & 27) downward only 0.001 (ea data point down by 0.001). These were only cavities to average 0.402 in size. Should measure steel to understand root cause!

Troubleshooting Flash Problems

Mold parting line shut-offs can be checked with PLASTIGAGE® clearance indicator. The green is PG-1 (aka SPG-1, MPG-1 & ZSPG-1, etc depending on supplier) for checking clearances of 0.001 to 0.003", the red is PR-1 for checking clearances of 0.002 to 0.006" ... there is also a blue (PB-1) for 0.004 - 0.009 and a yellow (PY-1) for 0.009 - 0.020 inch clearances. These wax strings can be laid across the parting line, then close mold with normal tonnage to check the actual clearance which may be causing flash. The wax strings are made such that the round string is compressed to a width based on clearance ... you compare the observed width with a scale included on packaging to determine clearance. This product can be purchased at suppliers to automotive engine rebuilders (auto parts store).

The best process solution to flash is to:
1. Reduce the injection pressure used for packing and/or filling. Sometimes fast filling, which likely uses high filling pressures to accomplish, will blow the mold open during filling. The mold may be flashed before the VP transfer to pack or hold takes place (VP – from velocity control to pressure control); set pack and hold pressures to zero to determine if flashing takes place during filling. If during fill, reduce velocity setpoints which should reduce required filling pressure. If flash occurs during pack, then reduce pressure as needed.

Other solutions include:
2. Increase clamp tonnage if available. Less tonnage may help if mold is small (less than 2/3 tie bar space) resulting in possible platen wrap around the mold which creates clearance below mid-section of mold.
3. Adjust VP position and/or method; transfer earlier.
4. Reduce the shot size and/or cushion size.
5. Reduce melt temperature.
6. Improve drying of hygroscopic resins. Hygroscopic resins which are processed when wet can experience large drops in molecular weight to the point that the resin flows very easily. Moisture bubbles also can act like a plasticizer by reducing melt density which also makes the resin flow more easily.
7. Perform initial packing at a very low pressure to set up skin at location of flash, then increase pressure with next stage (hold pressure) to pack part as needed.
8. Reduce melt residence time if resin is degrading.
9. Check parallelism of machine's platens.
10. Check injection delay or pressure switch for injection; verify clamp tonnage achieved before start of injection.
11. Reduce pressure for restart after cycle interrupt.
12. Non-Process: Repair mold parting line/shut-offs; check mold design for adequate support pillars.

Cavity PSI Transducer Calculations

In order to plan or select the transducer, we need to calculate the full scale pressure. The molding press typically wants to know what 10 volts equals in terms of pressure (in 1st calc below: 10 v = 30,859.6 psi) This is important to know so we don't create a scenario whereby the full scale pressure is 10X (or even 50X) what the machine can generate. A typical molding press can create 10K - 40K psi in cavity. This writer has seen transducer applications whereby small pins are used with large charge amps whereby the full scale pressure is greater than 1M psi; thus, resulting in poor resolution in output graphic displays ... this is corrected by using a smaller charge amp (with piezoelectric transducers) or a smaller maximum load transducer.

If using a piezoelectric type direct read pressure transducer (not behind an ejector pin). Determine transducer sensitivity value (pC/bar) and the charge amp full scale range (pC) ... use formula as follows:

$$\frac{range\ (pC)}{sensitivity\ (pC/bar)} = \frac{20,000\ pC}{\left(9.4\ \dfrac{pC}{bar} \times \dfrac{1\ bar}{14.504\ psi}\right)} = 30859.6\ psi$$

in this example above, the charge amp is 20,000 pC and transducer is 9.4 pC/bar (piezo transducer) as shown.

If using a piezoelectric type transducer below an ejector pin, then the formula is altered as follows to convert the force back to a pressure by dividing by the ejector pin area:

$$\frac{\dfrac{range\ (pC)}{sensitivity\ (pc/N)}}{area\ (in^2)} = \frac{\dfrac{20,000\ pC}{\left(4.4\ \dfrac{pC}{N} \times \dfrac{1\ N}{0.2248\ lbs}\right)}}{\left(\dfrac{\pi \times 0.0866\ inches^2}{4}\right)} = \frac{173,480\ lbs}{in^2} = psi$$

If using a strain gage transducer, divide transducer load rating by ejector pin area (there are often different transducer load ratings available):

$$\frac{125\ lbs}{\dfrac{\pi \times 0.062^2}{4}} = \frac{125\ lbs}{0.003019\ in^2} = 41,403.47\ psi$$

The machine would send 10v to transducer and receive an output back based on load...if machine received 2.65 volts back, it would compute pressure as follows:

$$\frac{10\ v}{41,403.47\ psi} = \frac{2.65\ v}{X\ psi}$$

$$\frac{2.65\ v \times 41,403.47\ psi}{10\ v} = 10,971.92\ psi$$

Miscellaneous Suppliers: Page 1

Ampco Metal, Inc.
P.O. Box 2004
1117 East Algonquin Rd
Arlington Heights, IL 60005
(800) 844-6008   (847) 437-6008 ...fax
www.ampcometal.com

Armoloy Corporation
118 Simonds Ave
Dekalb, IL 60115
(815) 758-6657   (815) 758-0268 ...fax

Oerlikon Balzer Coatings (TiN)
1181 Jansen Farm Ct.
Elgin, IL 60123
(847) 844-1753   (847) 695-5200
info.balzers.us@oerlikon.com

Bohler-Uddeholm Corporation
2505 Millenium Dr
Elgin, IL 60124
(800) 638-2520   (630) 883-3101
www.bucorp.com

Materion Brush Perf Alloy
606 Lamont Road
Elmhurst, IL 60125
(800) 323-2438 Sales
(630) 832-9650   (630) 832-9657 ...fax
www.brushwellman.com

Choice Mold Components, Inc.
44347 Reynolds Drive
Clinton Twp., MI 48036-1243
(800) 425-3171   (800) 425-3177 ...fax
www.choicemold.com

Crucible Industries LLC. (Steel)
575 State Fair Blvd.
Solvay, NY 13209
(800) 365-1180 x2
www.crucibleservice.com

DiamondBLACK
www.boroncarbidecoating.com
www.bodycote.com

Dicronite Dry Lube
800-874-4319
www.dicronite.com

D-M-E Company
World Headquarters
29111 Stephenson Hwy.
Madison Heights, MI 48071
(800) 626-6653
(248) 398-6000   (888) 808-4363 ...fax
www.dme.net

Edro Engineering
Corporate Headquarters
20500 Carrey Road
Walnut, Ca 91789
(800) 368-3376   (909) 594-2154...fax
www.edro.com

Electrolizing Inc.
20 Houghton Street
Providence, RI 02904-1014
(401) 861-5900   (401) 421-9512 ...fax
www.electrolizing.com

Ferro-Tic® SBC
41 Governor Dr
Newburgh, NY 12550
(845) 567-8300   (845) 567-0540 ...fax
www.ferro-tic.com

Hansen Couplings is now:
Eaton Hyd Group USA
1000 West Bagley Road
Berea, OH 44017
(440) 826-1115   (440) 826-1105 ...fax
www.eaton.com/hydraulics

Hasco America Inc.
270 Rutledge Road, Unit B
Fletcher, NC 28732
(800) 266-4272   (828) 650-2600
www.hasco.com
email: quotes.america@hasco.com

Husky Inj Molding Systems
500 Queen Street S.
Bolton, Ontario, L7E 5S5
Canada
www.husky.ca
(905) 951-5000

Husky Injection Molding Sys
288 North Road
Milton, VT 05468
(802) 859-8000   (802) 859-8499 ...fax

INCOE Corporation
1740 E. Maple Road
Troy, MI 48083
(248) 616-0220   (248) 616-0225
email: info@incoe.com

Lamina, Inc.
38505 Country Club Drive, Suite 100
Farmington Hills, MI.
U.S.A. 48331
(800) 652-6462
(248) 489-9122   (248) 553-6842 ...fax
email: sales@lamina.com

data verified 10.21.2011

Miscellaneous Suppliers: Page 2

Mold Base Industries, Inc.
7450 & 7501 Derry Street
Harrisburg, PA 17111
(800) 241-6656
(717) 564-2250 ...fax
www.moldbase.com
sales@moldbase.com

M.J.Vail Company, Inc.
10 Ilene Court, Suite 4
Hillsborough, NJ 08844
(800) 526-6003   In NJ- 908 359-6000
(908) 359-0518 ...fax
www.mjvail.com

Master Unit Die Products, Inc.
see DME

Mold-Masters Systems Inc.
233 Armstrong Avenue
Georgetown, ON L7G 4X5 Canada
(905) 877-0185   (905) 873-2818 ...fax
www.moldmasters.com
email: info@moldmasters.com

Mold-Masters Injectioneering LLC
103 Peyerk Court, Unit E
Romeo, MI 48065
(586) 752-6551   (586) 752-6552
email: mmi-inq@moldmasters.com

Mold-Tech (Textures)
34497 Kelly Road
Fraser, MI 48026
United States of America
(586) 296-5500   (586) 296-5691 ...fax
email : michigan@mold-tech.com
www.mold-tech.com

Mold-Tech (Textures)
5195 North Lake Dr
Lake City, GA 30260
(404) 363-6900   (404) 361-4105 ...fax
email: southest@mold-tech.com

Osco, Inc.
2937 Waterview Drive
Rochester Hills, MI 48309
(800) 499-6726
(248) 852-7310   (248) 852-7183 ...fax
email: sales@oscosystems.com

PCS Company
34488 Doreka Dr
Fraser, MI 48026
(800) 521-0546   (800) 505-3299 ...fax
www.pcs-company.com
email: sales@pcs-company.com

PENBERTHY, INC.
320 Locust Street
Prophetstown, IL 61277
(815) 537-2311   (815) 537-5764 ...fax
sales@pcc-penberthy.com
www.penberthy-online.com

Performance Alloys
N116 W18515 Morse Drive
Germantown, Wisconsin 53022
(262) 255-6662   (262) 255-3655 ...fax
(800) 272-3031
www.performancealloys.net

Progressive Components
235 Industrial Drive
Wauconda, IL 60084
(800) 269-6653   (800) 462-6653 ...fax
(847) 487-1000   (847) 487-5903 ...fax
email: sales@procomps.com
www.procomps.com

Parker Hannifin Corp Quick Coupling Division
Call for nearest distributor
(800) 272-7537

Poly-Plating, Inc. (Poly-Ond Coatings)
2096 Westover Road
Chicopee, MA 01022
(413) 593-5477   (413) 593-1631 ...fax
(941) 371-8555 ...Florida
www.poly-ond.com

(ROUNDMATE) Pleasant Precision Inc
13840 State Route 68
Kenton, OH 43326-9302
(866) 774-8479   (419) 675-3283 ...fax
email: info@roundmate.com

Ross Tool Corporation
1001 Chestnut St; PO Box 66
Warren, RI 02885
(800) 377-2603   (800) 258-8420 ...fax
(401) 245-1814   (401) 245-2250 ...fax

Ryerson Steel, Inc.
2621 W. 15th Place
Chicago, IL 60608
773-762-2121
email: questions@ryerson.com
www.ryerson.com

Synventive Molding Solutions
10 Centennial Drive
Peabody, MA 01960
(800) 367-5662
(978) 750-8065   (978) 646-3600
email: info@synventive.com

data verified 10.21.2011

Conversion Factors

LENGTH

1 inch = 25.4 mm
1 mm = 0.03937 in
1 foot = 30.48 cm
1 micron = 0.001 mm
1 micron = 0.0000394 in
1 inch = 2.54 cm
1 meter = 39.37 in
1 meter = 100 cm
1 microinch = 0.000001 in
1 microinch = 0.0254 microns
(printer's)
1 pica = 0.166 in
1 point = 0.01384 in

WEIGHT

1 lb = 453.6 gr
1 lb = 16 oz
1 gram = 0.035 oz
1 kg = 1000 gr
1 kg = 2.2046 lb
1 oz = 28.35 gr
1 metric ton = 2204.6 lb
1 metric ton = 1000 kg

ANGLES

1 degree = 0.01745 radian
1 degree = $\pi/180$ radian

VOLUME

1 cu in = 16.387 cc
1 cu ft = 1728 cu in
1 qt = 0.946 L
1 gal = 128 oz
1 cc = 1 gr (water)
1 gal = 8.33 lb
1 cu ft = 7.48 gal

AREA

1 sq in = 6.452 cc
1 sq ft = 144 sq in
1 acre = 43560 sq ft
1 sq cm = 0.155 sq in
1 sq ft = 0.111 sq yd
1 sq mm = 0.00155 sq in
1 sq km = 0.3861 sq mi

PRESSURE

1 in Hg = 13.6 in H_2O
1 kg/cm^2 = 14.223 psi
1 bar = 14.5 psi
1 atmos = 14.696 psi
1 MPa = 145 psi

ENERGY

1 BTU = 777.97 ft lb
1 cal = 3.09 ft lb
1 BTU = 252 cal
1 kwh = 3412 BTU
1 H.P. = 746 watts
1 ton (refrig) = 12000 Btu/hr
1 ton (refrig) = 3517 watts

SPECIFIC HEAT & HEAT TRANSFER

1 Cal/sec cm °C = 2903 BTU-in/hr ft^2 °F
1 Cal/sec cm °C = 241.9 BTU/hr ft °F
1 W/(m °K) = 0.0023884 Cal/sec cm °C
1 W/(m °K) = 0.5778 BTU/hr ft °F
1 W/(m °K) = 6.9335 BTU-in/hr ft^2 °F
1 Btu/(Lb °F) = 4.184 KJ/(Kg °K)
1 Btu/(Lb °F) = 4184 J/(Kg °K)
1 Cal/(g °C) = 1 Btu/(Lb °F)

TEMPERATURE CONVERSIONS

°C = (°F-32)/1.8
°F = (°C x 1.8) + 32
°K = (°F+459.67)/1.8

Trig Table for Right Triangles

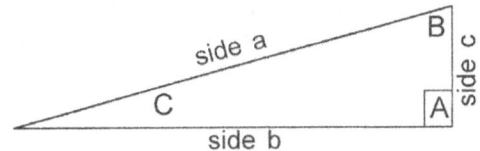

Find side	Formula	Find Angle	Formula
a	$\sqrt{b^2 + c^2}$	C	$\frac{\text{side c}}{\text{side a}} = \text{Sin C}$
a	$\frac{\text{side c}}{\text{Sin C}}$	C	$\text{Sin}^{-1}\left[\frac{\text{side c}}{\text{side a}}\right]$
a	$\frac{\text{side c}}{\text{Cosin B}}$	C	$\frac{\text{side b}}{\text{side a}} = \text{Cosin C}$
a	$\frac{\text{side b}}{\text{Sin B}}$	C	$\frac{\text{side c}}{\text{side b}} = \text{Tan C}$
a	$\frac{\text{side b}}{\text{Cosin C}}$	C	$\text{Tan}^{-1}\left[\frac{\text{side c}}{\text{side b}}\right]$
b	$\sqrt{a^2 - c^2}$	C	$\frac{\text{side b}}{\text{side c}} = \text{Cotan C}$
b	side a x Sin B	C	$\frac{\text{side a}}{\text{side b}} = \text{Secant C}$
b	side a x Cosin C	C	$\frac{\text{side a}}{\text{side c}} = \text{Cosec C}$
b	$\frac{\text{side c}}{\text{Tan C}}$	B	$\frac{\text{side b}}{\text{side a}} = \text{Sine B}$
b	side c x Tan B	B	$\frac{\text{side c}}{\text{side a}} = \text{Cosin B}$
c	$\sqrt{a^2 - b^2}$	B	$\frac{\text{side b}}{\text{side c}} = \text{Tan B}$
c	side b x Tan C	B	$\text{Tan}^{-1}\left[\frac{\text{side b}}{\text{side c}}\right]$
c	side a x Sin C	B	$\frac{\text{side c}}{\text{side b}} = \text{Cotan B}$
c	side a x Cosin B	B	$\frac{\text{side a}}{\text{side c}} = \text{Secant B}$
c	$\frac{\text{side b}}{\text{Tan B}}$	B	$\frac{\text{side a}}{\text{side b}} = \text{Cosec B}$

Tap, Number & Letter Drill Sizes

SIZE	T.P.I.	DRILL SIZE	DECIMAL EQUIV	SIZE	T.P.I.	DRILL SIZE	DECIMAL EQUIV	SIZE	T.P.I.	DRILL SIZE	DECIMAL EQUIV	SIZE	T.P.I.	DRILL SIZE	DECIMAL EQUIV
		80	0.0135			37	0.1040			C	0.2420			5/8	0.6250
		79	0.0145	6	32	36	0.1065			D	0.2460	11/16	24	41/64	0.6406
		1/64	0.0156			7/64	0.1094			1/4	0.2500	3/4	10	41/64	0.6406
		78	0.0160			35	0.1100			E	0.2500			21/32	0.6562
		77	0.0180			34	0.1110	5/16	18	F	0.2570	3/4	12	43/64	0.6719
		76	0.0200	6	40	33	0.1130			G	0.2610	3/4	16	11/16	0.6875
		75	0.0210			32	0.1160	5/16	20	17/64	0.2656	3/4	20	45/64	0.7031
		74	0.0225			31	0.1200			H	0.2660			23/32	0.7187
		73	0.0240			1/8	0.1250	5/16	24	I	0.2720	13/16	12	47/64	0.7344
		72	0.0250			30	0.1285			J	0.2770	13/16	16	3/4	0.7500
		71	0.0260	8	32	29	0.1360			K	0.2810	13/16	20	49/64	0.7656
		70	0.0280	8	36	29	0.1360	5/16	32	9/32	0.2812	7/8	9	49/64	0.7656
		69	0.0292			28	0.1405			L	0.2900			25/32	0.7812
		68	0.0310			9/64	0.1406			M	0.2950	7/8	12	51/64	0.7969
		1/32	0.0312			27	0.1440			19/64	0.2969	7/8	14	51/64	0.7969
		67	0.0320			26	0.1470			N	0.3020	7/8	16	13/16	0.8125
		66	0.0330	10	24	25	0.1495	3/8	16	5/16	0.3125	7/8	20	53/64	0.8281
		65	0.0350			24	0.1520			O	0.3160			27/32	0.8437
		64	0.0360			23	0.1540			P	0.3230	15/16	12	55/64	0.8594
		63	0.0370			5/32	0.1562	3/8	20	21/64	0.3281	15/16	16	7/8	0.8750
		62	0.0380	10	32	22	0.1570	3/8	24	Q	0.3320	15/16	20	57/64	0.8906
		61	0.0390			21	0.1590			R	0.3390	1	8	7/8	0.8750
		60	0.0400			20	0.1610	3/8	32	11/32	0.3438	1	12	59/64	0.9219
		59	0.0410			19	0.1660			S	0.3480	1	14	59/64	0.9219
		58	0.0420			18	0.1695			T	0.3580			15/16	0.9375
		57	0.0430	12	24	17	0.1730	7/16	14	U	0.3680	1	16	15/16	0.9375
0	80	56	0.0465			16	0.1770			3/8	0.3750	1	20	61/64	0.9531
		3/64	0.0469	12	28	15	0.1800			V	0.3770			31/32	0.9687
		55	0.0520			14	0.1820	7/16	20	W	0.3860	1-1/8	7	63/64	0.9844
1	64	54	0.0550			13	0.1850			25/64	0.3906			1	1.0000
1	72	53	0.0595			3/16	0.1875	7/16	24	X	0.3970	1-1/8	12	1-1/32	1.0312
		1/16	0.0625			12	0.1890	7/16	28	Y	0.4040			1-3/64	1.0469
		52	0.0635			11	0.1910			13/32	0.4062	1-1/8	16	1-1/16	1.0625
2	56	51	0.0670			10	0.1935			Z	0.4130	1-1/8	18	1-1/16	1.0625
2	64	50	0.0700			9	0.1960	1/2	13	27/64	0.4219				
		49	0.0730			8	0.1990			7/16	0.4375				
		48	0.0760	1/4	20	7	0.2010	1/2	20	29/64	0.4531				
3	48	5/64	0.0781			13/64	0.2031	1/2	24	29/64	0.4531				
		47	0.0785			6	0.2040	1/2	28	15/32	0.4688				
3	56	46	0.0810			5	0.2055	9/16	12	15/32	0.4688				
		45	0.0820	1/4	24	4	0.2090			31/64	0.4844				
		44	0.0860			3	0.2130	9/16	18	1/2	0.5000				
4	40	43	0.0890	1/4	28	7/32	0.2188	9/16	24	33/64	0.5156				
4	48	42	0.0935	1/4	32	7/32	0.2188	5/8	11	17/32	0.5312				
		3/32	0.0938			2	0.2210	5/8	12	35/64	0.5469				
		41	0.0960			1	0.2280	5/8	18	9/16	0.5625				
		40	0.0980			A	0.2340	5/8	24	37/64	0.5781				
5	40	39	0.0995			15/64	0.2344			19/32	0.5937				
5	44	38	0.1015			B	0.2380	11/16	12	39/64	0.6094				

PIPE DRILL SIZES (NPT)*

SIZE	T.P.I.	DRILL SIZE	DECIMAL EQUIV
1/16		D	0.2460
1/8	27	Q	0.3320
1/4	18	7/16	0.4375
3/8	18	9/16	0.5625
1/2	14	45/64	0.7031
3/4	14	29/32	0.9062
1	11½	1-9/64	1.1406
1¼	11½	1-31/64	1.4844
1½	11½	1-47/64	1.7344
2	11½	2-13/16	2.2031

* FOR TAPPING WITHOUT REAMING